AGENT INTELLIGENCE THROUGH DATA MINING

MULTIAGENT SYSTEMS, ARTIFICIAL SOCIETIES, AND SIMULATED ORGANIZATIONS

International Book Series

Series Editor: Gerhard Weiss, *Technische Universität München*

Books in the Series:

AGENT INTELLIGENCE THROUGH DATA MINING, Andreas Symeonidis, Pericles Mitkas, ISBN 0-387-24352-6

EXTENDING WEB SERVICES TECHNOLOGIES: ***The Use of Multi-Agent Approaches,*** edited by Lawrence Cavedon, Zakaria Maamar, David Martin, and Boualem Benatallah, ISBN 0-387-23343-1

AUTONOMY ORIENTED COMPUTING: ***From Problem Solving to Complex Systems Modeling,*** Jiming Liu, XiaoLong Jin, and Kwok Ching Tsui, ISBN1-4020-8121-9

METHODOLOGIES AND SOFTWARE ENGINEERING FOR AGENT SYSTEMS: ***The Agent-Oriented Software Engineering Handbook,*** **edited by** Federico Bergenti, Marie-Pierre Gleizes, Franco Zambonelli

AN APPLICATION SCIENCE FOR MULTI-AGENT SYSTEMS, edited by Thomas A. Wagner, ISBN: 1-4020-7867-6

DISTRIBUTED SENSOR NETWORKS, edited by Victor Lesser, Charles L. Ortiz, Jr., Milind Tambe, ISBN: 1-4020-7499-9

AGENT SUPPORTED COOPERATIVE WORK, edited by Yiming Ye, Elizabeth Churchill, ISBN: 1-4020-7404-2

AGENT AUTONOMY, edited by Henry Hexmoor, Cristiano Castelfranchi, Rino Falcone, ISBN: 1-4020-7402-6

REPUTATION IN ARTIFICIAL SOCIETIES: ***Social Beliefs for Social Order***, by Rosaria Conte, Mario Paolucci, ISBN: 1-4020-7186-8

GAME THEORY AND DECISION THEORY IN AGENT-BASED SYSTEMS, edited by Simon Parsons, Piotr Gmytrasiewicz, Michael Wooldridge, ISBN: 1-4020-7115-9

AGENT INTELLIGENCE THROUGH DATA MINING

by

Andreas L. Symeonidis
Pericles A. Mitkas
Aristotle University of Thessaloniki
Greece

Andreas L. Symeonidis
Postdoctoral Research Associate
Electrical and Computer Engineering Dept.
Aristotle University of Thessaloniki
54124,Thessaloniki, Greece

Pericles A. Mitkas
Associate Professor
Electrical and Computer Engineering Dept.
Aristotle University of Thessaloniki
54124,Thessaloniki, Greece

Library of Congress Cataloging-in-Publication Data

A C.I.P. Catalogue record for this book is available
from the Library of Congress.

AGENT INTELLIGENCE THROUGH DATA MINING
by Andreas L. Symeonidis and Pericles A. Mitkas
Aristotle University of Thessaloniki, Greece

Multiagent Systems, Artificial Societies, and Simulated Organizations Series
Volume 14

ISBN 978-1-4419-3724-7 e-ISBN 978-0-387-25757-0

Printed on acid-free paper.

9 8 7 6 5 4 3 2 1 SPIN 11374831, 11421214

springeronline.com

Andreas L. Symeonidis
dedicates this book to
his sister, Kyriaki,
and to the
"Hurri"cane of his life...

Pericles A. Mitkas
dedicates this book to
Sophia,
Alexander, and Danae.
For all the good years...

Contents

List of Figures

List of Tables

Foreword

The only wisdom we can hope to acquire
Is the wisdom of humility: humility is endless

T.S. Elliot

Data mining[1] (DM) is the process of finding previously unknown, profitable and useful patterns hidden in data, with no prior hypothesis. The objective of DM is to use discovered patterns to help explain current behavior or to predict future outcome. DM borrows concepts and techniques from several long-established disciplines, among them, Artificial Intelligence, Database Technology, Machine Learning and Statistics. The field of DM has, over the past fifteen years, produced a rich variety of algorithms that enable computers to *learn* from large datasets new relationships/knowledge.

DM has witnessed a considerable growth of interest over the last five years, which is a direct consequence of the rapid development of the information industry. Data is no longer a scarce resource; it is abundant and it exists, in most of the cases, in databases that are geographically distributed. Most recent advances in Internet and World Wide Web have opened the access to various databases and data resources and, at the same time, they induce many more new problems to make intelligent usage of all data that are both *available* and *relevant*. New methods for *intelligent* data analysis to extract relevant information are needed. The Information Society requires the development of new, more intelligent methods, tools, and theories for the discovering and modeling of relationships in huge amounts of consolidated data warehouses.

[1] Data mining is also known as Knowledge Discovery in Databases (KDD).

The goal of this book is to give a self-contained overview of a relatively young but important to be area of research that is receiving steadily increasing attention in the past years, that is the intersection of Agent Technology (AT) and Data Mining. This intersection is leading to considerable advancements in the area of information technologies and drawing an increasing attention of both the research and industrial communities. This book is a good example of this trend. It is the result of three years of intense work in the frame of Agent Academy, an EU-funded project. In this kind of projects a balance between research and development usually exists.

My initial experience with the Agent Academy frame was as a reviewer of the project. At first, my experience with the combined application of Multi-Agent Systems (MAS) technology to design architectures of DM, and the utilization of data mining and KDD to support learning tasks in MAS research was not easy and it took me a while to arrive to really appreciate the on-going work. During the project life I saw how these concepts were evolving and getting accepted in a wider agent community such as AgentCities.

Now, it is clear to me that a new direction in Information Technology research emerges from the combination of both research areas. This book will help the reader to discover new ways to interpret Artificial Intelligence and Multi-Agent Systems concepts. Authors make it clear that the utilization of DM to support machine learning tasks in MAS research confirms the fact that these two technologies are capable of mutual enrichment and that their joint use results in information systems with emergent properties.

ULISES CORTES
Barcelona, 2005

Preface

This book addresses the arguably challenging problem of generating intelligence from data and transferring it to a separate, possibly autonomous, software entity. In its generality, the definition, generation, and transfer of intelligence is a difficult task better left to God or, at least, the very best of the AI gurus. Our point of view, however, is more focused.

The main thesis of the book is that knowledge hidden in voluminous data repositories, which are routinely created and maintained by today's applications, can be extracted by data mining and provide the inference mechanisms or simply the behavior of agents and multi-agent systems. In other words, these knowledge nuggets constitute the building blocks of agent intelligence. Here, intelligence is defined loosely so as to encompass a wide range of implementations from fully deterministic decision trees to evolutionary and autonomous communities of agents. In many ways, intelligence manifests itself as efficiency. We argue that the two, otherwise diverse, technologies of data mining and intelligent agents can complement and benefit from each other, yielding more efficient solutions.

The dual process of knowledge discovery and intelligence infusion is equivalent to learning, better yet, teaching by experience. Indeed, existing application data (i.e., past transactions, decisions, data logs, agent actions, etc.) are filtered in an effort to distill the best, most successful, empirical rules and heuristics. The process can be applied initially to train 'dummy' agents and, as more data are gathered, it can be repeated periodically or on demand to further improve agent reasoning. The book considers the many facets of this process and presents an integrated methodology with several examples. Our perspective leans more towards agent-oriented software engineering (AOSE) than artificial intelligence (AI).

The methodology is adapted and applied at three distinct levels of knowledge acquisition: a) the application data level, b) the agent-behavior data level, and c) agent communities. Several existing data mining techniques are considered and some new ones are described. We also present a number of generic multi-agent models and then show how the process can be applied and validated. Three representative test cases, corresponding to the above three levels, are described in detail. The first one is a multi-agent system for order recommendations operating on top of a traditional ERP system of a retailer. Data from the ERP database are mined to develop the knowledge models for the various agent types. The second test case involves agents that operate in a corporate website and assist a user during his/her visit. The third system is a community of autonomous agents simulating an ecosystem with varying degrees of uncertainty.

Some of the more fundamental issues that this book aspires to tackle include the following:

1. Data mining technology has proven a successful gateway for discovering useful knowledge and for enhancing business intelligence in a range of application fields. Numerous approaches employ agents to streamline the process and improve its results. The opposite route of performing data mining for improving agent intelligence has not been often followed.

2. Incorporating knowledge extracted through data mining into already deployed applications is often impractical, since it requires reconfigurable software architectures, as well as human expert consulting. The coupling of data mining and agent technology, proposed within the context of this book, is expected to provide an efficient roadmap for developing highly reconfigurable software approaches that incorporate domain knowledge and provide decision making capabilities. The exploitation of this approach may considerably improve agent infrastructures, while also increasing reusability and minimizing customization costs.

3. The inductive nature of data mining imposes logic limitations and hinders the application of the extracted knowledge on deductive systems, such as multi-agent systems. The book presents a new approach that takes all the relevant limitations and considerations into account and provides a pathway for employing data mining techniques in order to augment agent intelligence.

4. Although bibliography on data mining and agent systems already abounds, there is no single, integrated approach for exploiting data

mining extracted knowledge. The reader has to go through a number of books, in order to get the "big picture". We expect that researchers and developers in the fields of software engineering, AI, knowledge discovery, and software agents will find this book useful.

According to an infamous piece attributed to Capes Shipping Agencies, Inc., Norfolk:

A captain is said to be a man who knows a great deal about very little and who goes along knowing more and more about less and less until finally he knows practically everything about nothing.

An engineer on the other hand is a man that knows very little about a great deal and keeps knowing less about more until he knows practically nothing about everything.

A [shipping] agent starts out knowing everything about everything but ends up knowing nothing about anything due mainly to his association with the captains and the engineers.

We would like to add that based on our methodology:

A software agent starts out knowing nothing about anything until it learns something about something and, by gradual training, ends up knowing more, but never everything, about this something.

Pericles A. Mitkas
Thessaloniki, January 2005

Acknowledgments

The motivation for writing this book stems from the successful outcome of a research project called Agent Academy and funded by the European Commission (IST-2000-31050). Pericles A. Mitkas was the general coordinator of Agent Academy and Andreas L. Symeonidis was heavily involved in the project from conceptualization to completion. The main objective of Agent Academy was the development of an integrated framework for constructing multi-agent applications and for improving agent intelligence by the exploitation of data mining techniques. Within the context of this project, the idea of embedding knowledge models, extracted via data mining, into agents and multi-agent systems has matured enough to become the book you are now holding. The authors would like to express their gratitude to all the members of the Agent Academy consortium, for their commitment to the project and their hard work.

PART I

CONCEPTS AND TECHNIQUES

Chapter 1

INTRODUCTION

1. The Quest for Knowledge

Early computers were designed, mainly, for number crunching. As memory became more affordable, we started collecting *data* at increasing rates. Data manipulation produced *information* through an astonishing variety of intelligent systems and applications. As data continued to amass and yield more information, another level of distillation was added to produce *knowledge*. Knowledge is the essence of information and comes in many flavors. Expert systems, knowledge bases, decision support systems, machine learning, autonomous systems, and intelligent agents are some of the many packages researchers have invented in order to describe applications that mimic part of the human mental capabilities. A highly successful and widely popular process to extract knowledge from mountains of data is *data mining.*

The application domain of **Data Mining** (DM) and its related techniques and technologies have been greatly expanded in the last few years. The development of automated data collection tools and the ensuing tremendous data explosion have fueled the imperative need for better interpretation and exploitation of massive data volumes. The continuous improvement of hardware along with the existence of supporting algorithms has enabled the development and flourishing of sophisticated DM methodologies. Issues concerning data normalization, algorithm complexity and scalability, result validation and comprehension have been successfully dealt with [Adriaans and Zantinge, 1996; Witten and Frank, 2000; Han and Kamber, 2001]. Numerous approaches have been adopted for the realization of autonomous and versatile DM tools to support all the appropriate pre- and post-processing steps of the knowledge discovery process in databases [Fayyad et al., 1996; Chen et al., 1996].

2. Problem Description

Since DM systems encompass a number of discrete, nevertheless dependent tasks, they can be viewed as networks of autonomous, yet collaborating units, which regulate, control and organize all the, potentially distributed, activities involved in the knowledge discovery process. Software agents, considered by many the evolution of objects, are autonomous entities that can perform these activities.

Agent technology has introduced a windfall of novel computer-based services that promise to dramatically affect the way humans interact with computers. The use of agents may transform computers into personal collaborators that can provide active assistance and even take the initiative in decision-making processes on behalf of their masters. Agents participate routinely in electronic auctions and roam the web searching for knowledge nuggets. They can facilitate "smart" solutions in small and medium enterprises in the areas of management, resource allocation, and remote administration. Enterprises can benefit immensely by expanding their strategic knowledge and weaponry.

Research on software agents has demonstrated that complex problems, which require the synergy of a number of distributed elements for their solution, can be efficiently implemented as a multi-agent system (MAS) [Ferber, 1999]. As a result, multi-agent technology has been repeatedly adopted as a powerful paradigm for developing DM systems [Stolfo et al., 1997; Kargupta et al., 1996; Zhang et al., 2003; Mohammadian, 2004].

In a MAS realizing a DM system, all requirements collected by the user and all the appropriate tasks are perceived as distinguished roles of separate agents, acting in close collaboration. All agents participating in a MAS communicate with each other by exchanging messages, encoded in a specific agent communication language. Each agent in the MAS is designated to manipulate the content of the incoming messages and take specific actions/decisions that conform to the particular reasoning mechanism specified by DM primitives.

Considerable effort is expended to formulate improved knowledge models for data mining agents, which are expected to operate in a more efficient and intelligent way. Moving towards the opposite direction (see Figure 1.1), we can envision the application of data mining techniques for the extraction of knowledge models that will be embedded into agents operating in diverse environments.

The interesting, non-trivial, implicit and potentially useful knowledge extracted by the use of DM [Fayyad et al., 1996] would be expected to

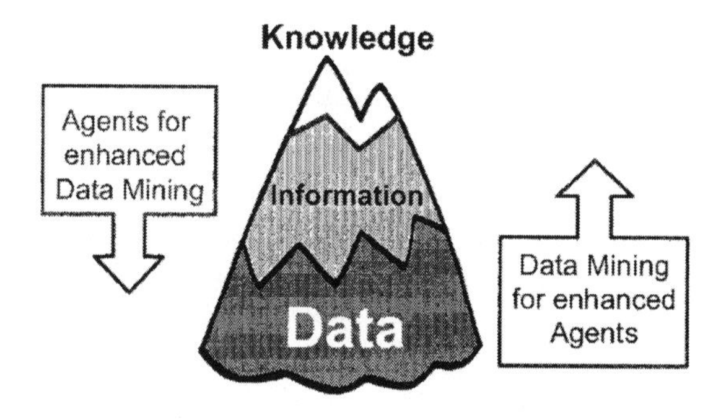

Figure 1.1. Mining for intelligence

find fast application on the development and realization of intelligence in agent technology (AT). The incorporation of knowledge based on previous observations may considerably improve agent infrastructures while also increasing reusability and minimizing customization costs. Unfortunately, limitations related to the nature of different types of logic adopted by DM and AT (inductive and deductive, respectively), hinder the unflustered application of knowledge to agent reasoning. If these limitations are overcome, then the coupling of DM and AT may become feasible.

3. Related Bibliography

A review of the literature reveals several attempts to couple DM and AT. Galitsky and Pampapathi [Galitsky and Pampapathi, 2003] in their work combine inductive and deductive reasoning, in order to model and process the claims of unsatisfied customers. Deduction is used for describing the behaviors of agents (humans or companies), for which we have complete information, while induction is used to predict the behavior of agents, whose actions are uncertain to us. A more theoretical approach on the way DM-extracted knowledge can contribute to AT performance has been presented by Fernandes [Fernandes, 2000]. In this work, the notions of data, information, and knowledge are modeled in purely logical terms, in an effort to integrate inductive and deductive reasoning into one inference engine. Kero et al. [Kero et al., 1995], finally, propose a DM model that utilizes both inductive and deductive components. Within the context of their work, they model the discovery of knowledge as an iteration between high level, user-specified patterns and their elaboration to (deductive) database queries, whereas they define the notion of a meta-query that performs the (inductive) analysis of these queries and their transformation to modified, ready-to-use knowledge.

In rudimentary applications, agent intelligence is based on relatively simple rules, which can be easily deduced or induced, compensating for the higher development and maintenance costs. In more elaborate environments, however, where both requirements and agent behaviors need constant modification in real time, these approaches prove insufficient, since they cannot accommodate the dynamic transfer of DM results into the agents. To enable the incorporation of dynamic, complex, and reusable rules in multi-agent applications, a systematic approach must be adopted.

4. Scope of the Book

Existing agent-based solutions can be classified according to the granularity of the agent system and inference mechanism of the agents. As shown in Figure 1.2, which attempts a qualitative representation of the MAS space, agent reasoning may fall under four major categories ranging from simple rule heuristics to self-organizing systems. Inductive logic and self-organization form two manifestations of data mining. Therefore, the shaded region delineates the area of interest of this book.

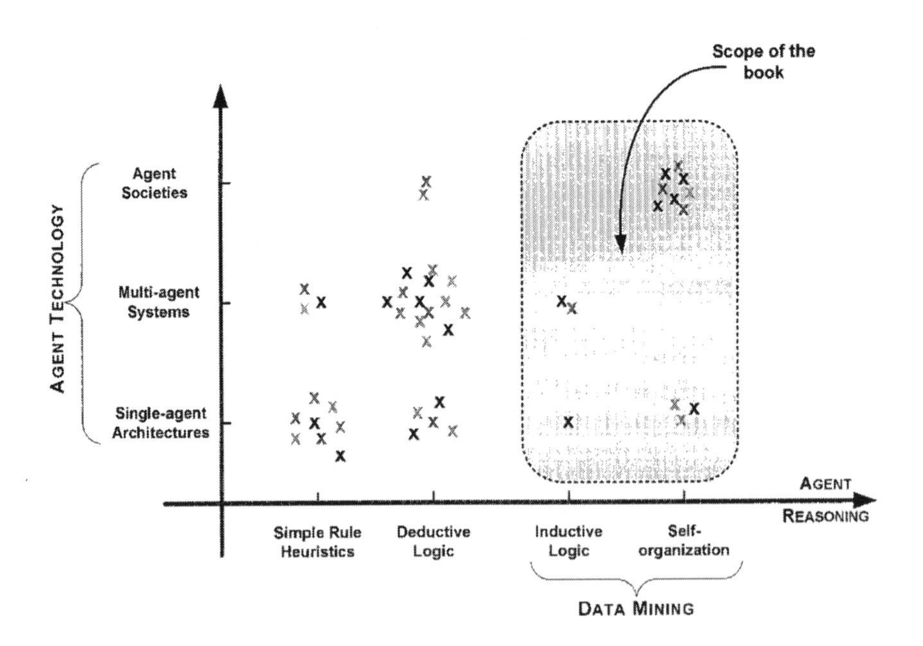

Figure 1.2. Agent-based applications and inference mechanisms

An agent is generated by a user, who may be a real or a virtual entity. This child agent can be a) created with enough initial intelligence, b) pre-trained to an acceptable competence level, or c) sent out untrained to learn on its own. We believe that intelligence should not be hard-coded in an agent because this option reduces agent flexibility and puts a heavy burden on the programmer's shoulders. On the other hand, ill-equipped agents are seldom effective. Taking the middle ground, a rather simple (dummy) agent can be created and then trained to learn, adapt, and get smarter and more efficient. This training session can be very productive, if information regarding other agents' experience and user preferences is available.

The embedded intelligence in an agent should be acquired from its experience of former transactions with humans and other agents that work on behalf of a human or an enterprise. Hence, an agent that is capable of *learning* can increase significantly its effectiveness as a personal collaborator and yield a reduction of workload for human users. The learning process is a non-trivial task that can be facilitated by extracting knowledge from the experience of other agents.

In this book we present a unified methodology for transferring DM-extracted knowledge into newly-created agents. Data mining is used to generate knowledge models which can be dynamically embedded into the agents. As new data accumulate, the process can be repeated and the decision structures can be updated, effectively retraining the agents. Consequently, the process is suitable for either upgrading an existing, non agent-based application by adding agents to it, or for improving the already operating agents of an agent-based application. The methodology relies heavily on the inductive nature of data mining, while taking into account its limitations.

In our approach, we consider three distinct types of knowledge, which correspond to different data sources and mining techniques: a) knowledge extracted by performing DM on historical datasets recording the business logic (at a macroscopic level) of a certain application, b) knowledge extracted by performing DM on log files recording the behavior of the agents (at a microscopic level) in an agent-based application, and c) knowledge extracted by the use of evolutionary data DM techniques in agent communities. These three types demarcate also three different modes of knowledge diffusion, which are defined in the book and demonstrated by three test cases.

5. Contents of the Book

The book is organized into nine chapters in addition to the current introduction.

Chapter 2 is a brief overview of data mining and knowledge discovery with an emphasis on the issues mentioned later in the book. We describe the basic steps of the knowledge discovery process and some classic methods for data preprocessing. The data mining techniques of classification, clustering, association rule extraction and genetic algorithms, which are used in subsequent chapters, are defined along with representative algorithms and short examples.

Chapter 3 introduces the reader to the basic concepts of software agents and agent intelligence. The functionality and key characteristics of agents are presented and juxtaposed to those of objects and traditional expert systems. We then define multi-agent systems and agent communities and discuss a number of agent communication issues.

Chapter 4 serves as the initiation to the idea of exploiting data mining results for improving agent intelligence. After the difference between deductive and inductive logic is established, we argue that a seamless marriage of the two logic paradigms, with each one contributing its strengths, may yield efficient agent-based systems with a good knowledge of the application domain. In this chapter we define the concepts of agent training, retraining, and knowledge diffusion and delineate three levels of knowledge transfer from data to agents. The chapter concludes with a review of software platforms for MAS development.

Chapter 5 comprises the presentation of the methodology for coupling data mining and agent technology. The methodology is first described in a unified manner and then adapted to the three levels of agent training. In each case, the steps of the process are discussed in more detail. The second half of the chapter is devoted to the Data Miner, an open-source platform that we have developed for supporting and automating a large part of the mechanism.

In order to demonstrate the feasibility of our approach, we have selected three different domains that correspond to the three levels of knowledge diffusion. For each domain, we have developed a framework that includes one or more multi-agent systems, new and existing DM techniques, and a demonstrator. Each framework can be generalized for a class of similar applications. Chapters 6 to 8 describe these three frameworks. In each case, the presentation includes analysis of the problem space, description of the MAS architecture and the DM experiments, results obtained with real data, and a discussion of benefits and drawbacks.

Chapter 6 presents an intelligent recommendation framework for business environments. An agent-based recommendation engine can be built on top of an operating ERP system and tap the wealth of data stored in the latter's databases. Such an approach can combine the decision support capabilities of more traditional approaches for supply chain management (SCM), customer relationship management (CRM), and supplier relationship management (SRM).

The MAS development framework, presented in Chapter 7, addresses the problem of predicting the future behavior of agents based on their past actions/decisions. Here we show how DM, performed on agent behavior datasets, can yield usable behavior profiles. We introduce κ-profile, a DM process to produce recommendations based on aggregate action profiles. The demonstrator in this case is a web navigation engine, which tracks user actions in large corporate sites and suggests possibly interesting sites. The framework can be extended to cover a large variety of web services and/or intranet applications.

Chapter 8 focuses on a typical example of knowledge diffusion in evolutionary systems. An agent community is used to simulate an ecosystem, where agents, representing living organisms, live, explore, feed, multiply, and eventually die in an environment with varying degrees of uncertainty. Genetic algorithms and agent communication primitives are exploited to implement knowledge transfer, which is essential for the survival of the community. The contents of this chapter include the description of a development platform for agent-oriented ecosystems, the formal model, as well as experimental results.

Chapter 9 is a first-order treatment of the agent retraining issue and a formal model is developed. Retraining efficiency is clearly dependent on the type and volume of available datasets. Experimental results for a few test cases are provided.

Finally, Chapter 10 takes a look at possible extensions of the methodology and outlines several additional areas of application, including environmental systems, e-auctions and enhanced software processing.

6. How to Read this Book

Every author believes that his book must be read from the first to the last page, preferably in this order. If you are not the traditional reader and/or have considerable background in the areas of data mining or agents, you may find the diagram in Figure 1.3 useful. Readers familiar with the knowledge discovery process may skip Chapter 2, while those working in agent-related fields may do the same with Chapter 3. We believe that everybody must read Chapters 4 and 5 because they lay out the main thesis of this book. The three application areas described in

Chapters 6, 7, and 8 should enhance the reader's understanding of the methodology and help clarify several issues. They could also be used as a roadmap for developing agent-based applications in three rather broad areas. These chapters can be read in any order depending on the reader's preferences or research interests. Having gone at least through the material in Chapters 6 and 7, the reader must have recognized the need and benefits of agent retraining and should be ready for Chapter 9. The last chapter can be read at any stage, even after this introduction, but its contents will be clearer to those of you who can wait till the end.

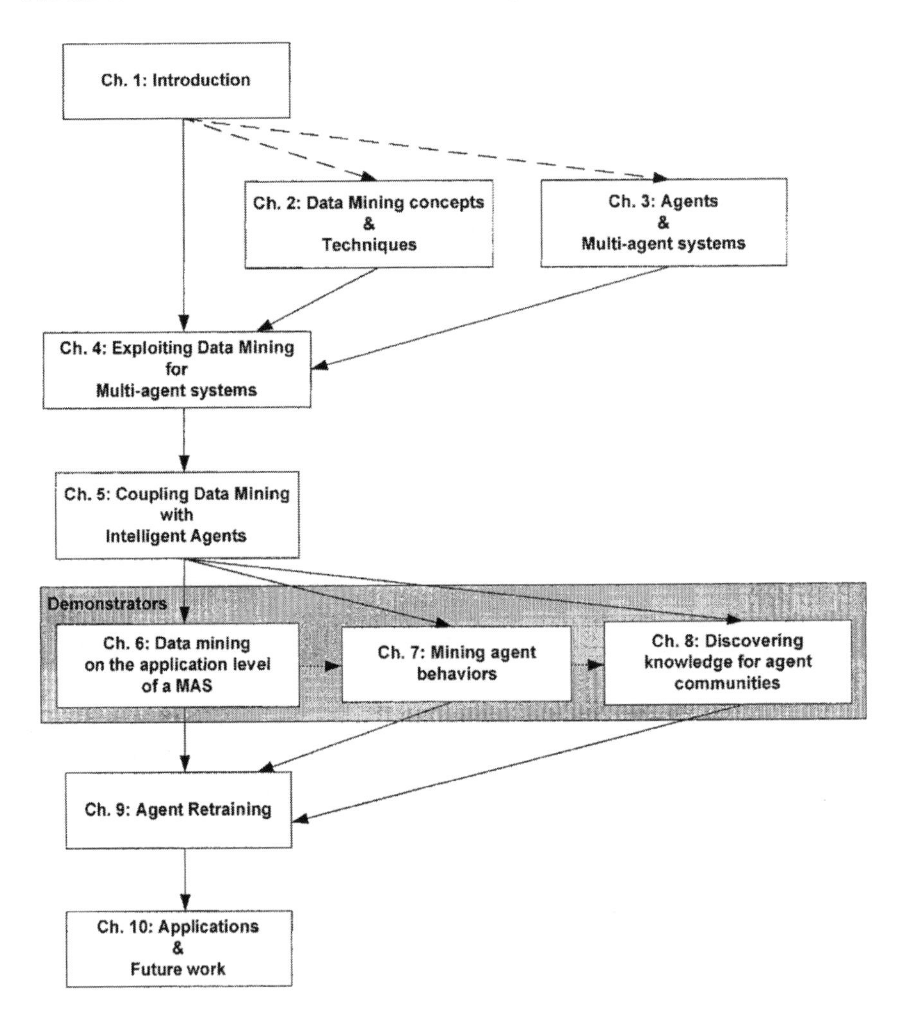

Figure 1.3. Alternative routes for reading this book

Chapter 2

DATA MINING AND KNOWLEDGE DISCOVERY: A BRIEF OVERVIEW

1. History and Motivation

1.1 The Emergence of Data Mining

Data Mining has evolved into a mainstream technology because of two complementary, yet antagonistic phenomena: a) the data deluge, fueled by the maturing of database technology and the development of advanced automated data collection tools and, b) the starvation for knowledge, defined as the need to filter and interpret all these massive data volumes stored in databases, data warehouses and other information repositories. DM can be thought of as the logical succession to Information Technology (IT). Considering IT evolution over the past 50 years (Figure 2.1), the first radical step was taken in the 60's with the implementation of data collection, while in the 70's, the first Relational Database Management Systems (RDBMS) were developed.

During the 80's, enhanced data access techniques began to emerge, the relational model was widely applied, and suitable programming languages were developed [Bigus, 1996].

Shortly (the 90's), another significant step in data management followed. The development of Data Warehouses (DW) and Decision Support Systems (DSS) allowed the manipulation of data coming from heterogeneous sources and supported multiple-level dynamic and summarizing data analysis.

Though the enhancements provided by DSS and the efficiency of DW are impressive, they alone cannot provide a satisfactory solution to solve the *data-rich but information-poor* problem, which requires advanced data analysis tools [Information Discovery Inc., 1999]. The human quest

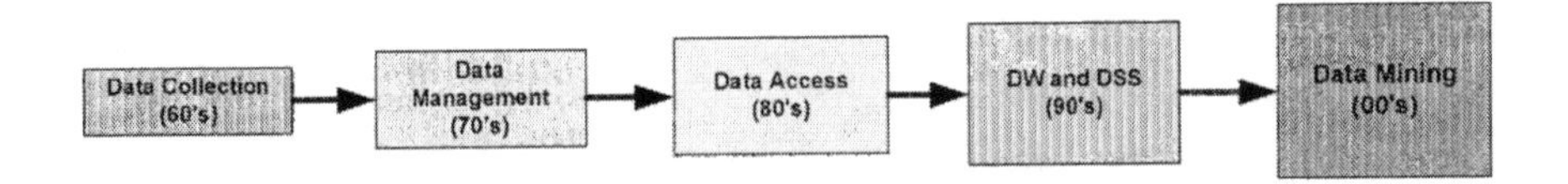

Figure 2.1. Technology evolution towards Data Mining

for knowledge and the inability to perceive the –continuously increasing– data volumes of a system has led to what we today call ***Data Mining***.

Figure 2.1 and Table 2.1 summarize the steps towards Data Mining, their enabling technologies and their fundamental characteristics (adapted from [Pilot Software Inc., 1999]).

Table 2.1. Steps in the evolution of Data Mining

Evolutionary Step	Enabling Technologies	Characteristics
Data Collection (60's)	Computers, tapes, disks	Retrospective, static data delivery
Data Management (70's)	DBMS, RDBMS	Dynamic data management at record level
Data Access (80's)	RDBMS, Structured Query Language (SQL), ODBC	Retrospective, dynamic data delivery at record level
DW & DSS (90's)	On-line analytical processing (OLAP), DW, multidimensional databases	Retrospective, dynamic data delivery at multiple levels
Data Mining (00's)	Advanced algorithms, multiprocessor computers massive datasets	Prospective, proactive information discovery

Clearly, DM emerged when the volumes of accumulated information had by far exceeded the quantities that a user could interpret. DM flourished rapidly, since:

a. In contrast to DSS, DM techniques are *computer-driven*, therefore can be fully automated.

b. DM solves the query formulation problem. After all, how could anyone access a database successfully, when the compilation of a structured query is not possible?

c. Finally, DM confronts the visualization and understanding of large data sets efficiently.

The above three factors, coupled with the rapid development of new and improved databases have made DM technology nowadays an integral part of information systems [Fayyad, 1996].

1.2 So, what is Data Mining?

Data Mining is closely related to *Knowledge Discovery in Databases* (KDD) and quite often these two processes are considered equivalent. Widely accepted definitions for KDD and DM have been provided by Fayyad, Piatetsky-Shapiro, & Smyth:

> *Knowledge Discovery in Databases is the process of extracting interesting, non-trivial, implicit, previously unknown and potentially useful information or patterns from data in large databases.*[Fayyad et al., 1996]

By the term "pattern" we define a model that is extracted from the data and assigns a number of common characteristics to them, whereas by the term "process" we stress the fact that KDD comprises many steps, such as data preprocessing, patterns query, and result validation.

> *Data Mining is the most important step in the KDD process and involves the application of data analysis and discovery algorithms that, under acceptable computational efficiency limitations, produce a particular enumeration of patterns over the data.*

Caution must be exercised in order to avoid confusion with other related but dissimilar technologies, such as data/pattern analysis, business intelligence, information harvesting, and data archeology.

Based on the above definitions, DM would be but one step of the KDD process. Nevertheless, within the context of this book, these two terms are used interchangeably, since the DM and KDD are viewed as two facets of the same process: the extraction of useful knowledge, in order to enhance the intelligence of software agents and multi-agent systems.

1.3 The KDD Process

The KDD process entails the application of one or more DM techniques to a dataset, in order to extract specific patterns and to evaluate them on the data. KDD is iterative and interactive, and comprises the following steps, shown schematically in Figure 2.2 [Han and Kamber, 2001; Xingdong, 1995].

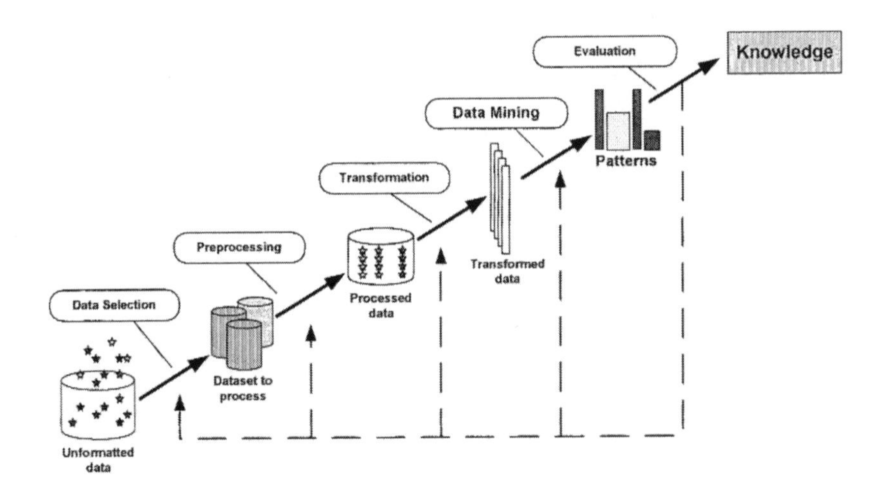

Figure 2.2. A schematic representation of the KDD process

1. **Identify the goal of the KDD process**
   Develop an understanding of the application domain and the relevant prior knowledge.

2. **Create a target data set**
   Select a data set, or focus on a subset of variables or data samples, on which discovery will be performed.

3. **Clean and preprocess data**
   Remove noise, handle missing data fields, account for time sequence information and known changes.

4. **Reduce and project data**
   Find useful features to represent the data, depending on the goal of the task.

5. **Identify a data mining method**
   Match the goals of the KDD process to a particular data mining method: e.g. summarization, classification, regression, clustering, etc.

6. **Choose a data mining algorithm**
   Select method(s) to be used for searching for patterns in the data.

7. **Apply data mining**

8. **Evaluate data mining results**
   Interpret mined patterns, possibly return to steps 1-7 for further iteration.

9. **Consolidate discovered knowledge**
   Incorporate this knowledge into another system for further action, or simply document it and report it to interested parties.

1.4 Organizing Data Mining Techniques

Data Mining can be applied to the vast majority of data organization schemes, most popular of which are the relational databases and data warehouses, as well as various transactional databases. Data mining is also used on a variety of advanced databases and information repositories, such as the object-oriented databases, spatial and temporal databases, text and multimedia databases, heterogeneous and legacy databases, and, finally, genomics databases and the world wide web (www) databases. For each database category, appropriate DM techniques and algorithms have been developed, to ensure an optimal result.

The outcome of DM can vary and is specified by the user in each case. In general, two are the main reasons for performing DM on a dataset: a) *Validation* of a hypothesis and, b) *Discovery* of new patterns. Discovery can be further divided into *Prediction*, where the knowledge extracted aims to better forecast the values of the entities represented in the dataset, and in *Description*, where the extracted knowledge aspires to improve the comprehension of the patterns discovered [Han and Kamber, 2001].

The most popular DM techniques, which *predict* and *describe* are [Fayyad et al., 1996]:

1. **Classification**
   The discovery of a knowledge model that classifies new data into one of the existing pre-specified *classes*.

2. **Characterization and Discrimination**
   The discovery of a valid description for a part of the dataset.

3. **Clustering**
   The identification of a finite number of *clusters* that group data based on their similarities and differences.

4. **Association-Correlation**
   The extraction of *association rules*, which indicate cause-effect relations between the attributes of a dataset.

5. **Outlier analysis**
   The identification and exclusion of data that do not abide by the behavior of the rest of the data records.

6. **Trend and evolution analysis**
   The discovery of trends and diversions and the study of the evolution of an initial state/hypothesis throughout the course of time.

The application of DM on a dataset may lead to the discovery of a great number of patterns. Nevertheless, not all of them are interesting:

> A pattern is considered to be *interesting* if it is easily understood by humans, valid on new or test data with some degree of certainty, potentially useful, novel, or if it validates some hypothesis that a user seeks to confirm [Piatetsky-Shapiro, 1991].

Interestingness may be objective or subjective, depending on whether it is based on statistical properties and the structure of the discovered patterns, or it is based on the user's belief in the data [Liu et al., 2001].

In order for the reader to understand the meaning of a subjectively interesting and not interesting pattern, we provide an example: Let us assume the database of a commercial store that stores information on the customers, suppliers, products, and orders. Table "*Customers*" contains, among others, columns "*Country*", "*City*", and "*Postal Code*". After having applied DM, a pattern of the form:

$$\text{When "}City : Berlin\text{" then "}Country : Germany\text{"}$$

is not interesting, since knowledge provided is obvious and not novel. One the other hand, a pattern of the form:

$$\text{When "}Product : Flight\ Wingman\ 2\text{" then "}Product : Cockpit\text{"}$$

is interesting, novel and useful for the store managers.

DM entails the confluence of several disciplines (Figure 2.3) and the degree of participation of these disciplines into DM delineates the different types of DM systems. We can classify these systems in various ways, depending on the criteria used. Some of these criteria are [Han and Kamber, 2001; Witten and Frank, 2000]:

- **The DM technique employed**
  DM systems can be classified either by the degree of user involvement (autonomous systems, query-driven systems), or by the data analysis technique utilized (database-oriented, OLAP, machine learning, statistics, etc.).

- **The type of DM-extracted knowledge**
  DM systems can be categorized as classification, characterization and

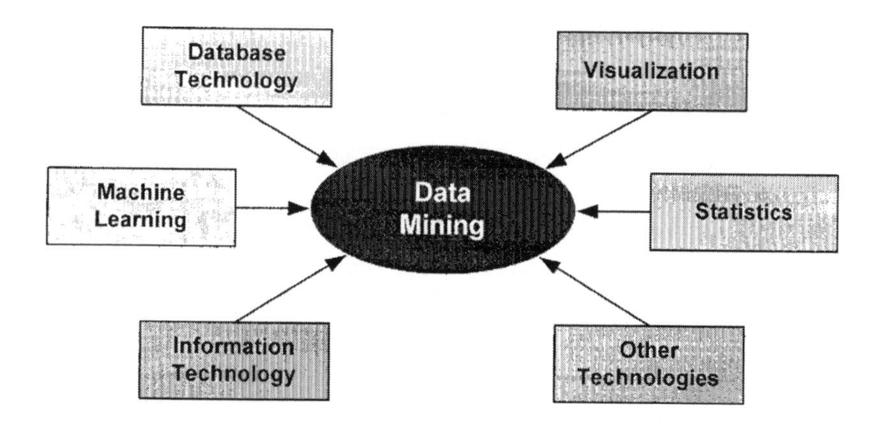

Figure 2.3. The confluence of different technologies into DM

discrimination, clustering, association-correlation, outlier discovery, and trend (and evolution) analysis systems.

- **The DW structure DM will be applied on**
  DM systems can be classified according to the type of data they will be applied on, or according to the DW underlying structure (transactional, spatial, temporal, genomics databases etc.)

- **The application domain DM-extracted knowledge is related to**
  DM systems can be classified according to their domain of application (financial, genetics, etc.)

In this book, we have used the last criterion, where the application domain is agent technology and multi-agent systems, to classify DM systems into three categories:

1) Systems that perform DM on the application level of agents

2) Systems that perform DM on the behavior level of agents

3) DM systems for evolutionary agent communities

In the remainder of this chapter, we outline the main data preprocessing methods, moving to the discussion of the four DM techniques used in subsequent chapters (classification, clustering, association rule extraction and genetic algorithms), as well as the most representative algorithms.

2. Data Preprocessing

2.1 The Scope of Data Preprocessing

In order to achieve the maximum benefit from the application of a DM algorithm on a dataset, preprocessing of the data is necessary, to ensure data integrity and validity. Basic preprocessing tasks include cleaning, transformation, integration, reduction, and discretization of the data [Kennedy et al., 1998; Pyle, 1999]. A brief overview of these tasks follows.

2.2 Data Cleaning

Real-world data are usually noisy and incomplete, and their cleaning includes all the processes of filling in missing values, smoothing out noise, and discovering outliers.

Missing values may be due to: a) equipment malfunction (i.e., the database is down), b) inconsistencies between data within a dataset and, thus, deletion, or c) omission, in case data are not understood or considered trivial [Friedman, 1977].

The most popular practices for handling missing values, are:

1. Ignore the whole dataset tuple (applied when the class attribute is missing – supervised learning)

2. Fill in the missing data manually (not applicable in the case of many tuples with missing values)

3. Use a special character denoting a missing value, i.e., "?"

4. Use the attribute mean to fill in the missing value

5. Use the attribute mean of all tuples belonging to the same class to fill in the missing value (supervised learning)

6. Use the most probable value to fill in the missing value

The most common reasons that introduce noise to data are: a) problems during the phases of data collection, entry or transmission, b) faulty instruments and technology limitations and, c) inconsistency in attribute naming conventions and existence of duplicate records.

Popular techniques for noise smoothing include binning, clustering, coordinated human-computer inspection, and regression. Most of these techniques include a data discretization phase, which is discussed later on in this Chapter.

2.3 Data Integration

This step entails the integration of multiple databases, data cubes, or files into a unified schema. Data integration can be accomplished at three levels [Han and Kamber, 2001]:

a. **Integration of the data store schema**
The goal is the integration of metadata from different sources and the solution of the "entity identification problem", i.e., how to identify identical or equivalent entities from multiple data sources.

b. **Detection and resolution of data value conflicts**
The goal is to handle cases where attribute values for the same real world entity, provided by different sources, are not the same, due to different representations or different scales, e.g., metric vs. British units.

c. **Management of redundant data**
The goal here is to identify and eliminate multiple copies of the same item, since the same attribute is often named differently in different databases (e.g. $A.cust\_id = B.cust\_id$). Through correlation analysis, the handling of redundant data is feasible.

2.4 Data Transformation

Another important preprocessing task is data transformation. The most common transformation techniques are:

- **Smoothing**, which removes noise from data.
- **Aggregation**, which summarizes data and constructs data cubes.
- **Generalization**, which is also known as concept hierarchy climbing.
- **Attribute/feature construction**, which composes new attributes from the given ones.
- **Normalization**, which scales the data within a small, specified range. The most dominant normalization techniques according to Weiss and Indurkhya are [Weiss and Indurkhya, 1998]:

  1) *min-max* normalization: Linear transformation is applied on the data. Let $\min_A$ be the minimum and $\max_A$ the maximum values of attribute A. $min - max$ normalization maps the original attribute A value ν to a new value ν' that lies in the [$new\_\min_A$,

$new\_max_A]$, according to Eq. 2.1:

$$\nu' = \frac{\nu - \min_A}{\max_A - \min_A}(new\_\max{}_A - new\_\min{}_A) + \\ + \; new\_\min{}_A \quad (2.1)$$

2) *z-score* normalization: Attribute A is normalized with respect to its average value and standard deviation (Eq. 2.2):

$$\nu' = \frac{\nu - mean_A}{std_A} \quad (2.2)$$

3) *decimal scaling* normalization: The values of attribute A are normalized by shifting their decimal part (Eq. 2.3):

$$\nu' = \frac{\nu}{10^j} \quad (2.3)$$

where j is the smallest integer that satisfies $\max(|\nu'|) < 1$.

2.5 Data Reduction

Data reduction techniques are used in order to obtain a new representation of the data set that is much smaller in volume, but yet produces the same (or almost the same) analytical results. The most common reduction strategies are a) data cube aggregation, b) dimensionality reduction, c) numerocity reduction, and d) discretization and concept hierarchy generation [Barbará et al., 1997].

2.6 Data Discretization

As already mentioned, data discretization is a strategy for data reduction, but with particular importance. Discretization methods are applied both on numeric and categorical data. Specific methods include:

- Discretization methods for numeric data [Kerber, 1992]
  - Binning
  - Histogram Analysis
  - Cluster Analysis
  - Entropy-Based Discretization
  - Segmentation by natural partitioning
- Discretization methods for categorical data [Han and Fu, 1994]
  - Specification of a partial ordering of attributes explicitly at the schema level by users or experts

- Specification of a portion of a hierarchy by explicit data grouping
- Specification of a set of attributes, but not of their partial ordering

3. Classification and Prediction

3.1 Defining Classification

We begin the discussion of four major data mining techniques with classification. Classification is a supervised two-step process. During the first step, a segment of the dataset, *the training set*, is used to extract an accurate model that maps the data of the training set into user predefined *classes* (groups). During the second step, the model is used either to classify any new data tuple or dataset introduced, or to extract more accurate classification rules. Classification could be defined as:

> *Classification is the process that finds the common properties among a set of objects in a dataset and classifies them into different classes, according to a classification model.* [Chen et al., 1996]

The core requirements for successful classification are a well-defined set of classes and a training set of pre-classified examples [Han and Kamber, 2001].

The most popular classification techniques are: a) Bayesian classification, b) decision trees, and c) neural networks. In the rest of this section we briefly describe the first two.

3.2 Bayesian Classification

As its name implies, Bayesian classification attempts to assign a sample x to one of the given classes $c_1, c_2, .., c_N$, using a probability model defined according to the Bayes theorem. The latter calculates the *posterior* probability of an event, conditional on some other event [Bretthorst, 1994].

Basic prerequisites for the application of bayesian classification are [Weiss and Kulikowski, 1991]:

1. Knowledge of the prior probability for each class c_i.
2. Knowledge of the conditional probability density function for $p(x|c_i) \in [0, 1]$

It is then possible to calculate the *posterior* probability $p(c_i|x)$ using the Bayes formula:

$$q(c_i|x) = \frac{p(x|c_i)\, p(c_i)}{p(x)} \tag{2.4}$$

where $p(x)$ is the prior probability of x.

Each new data tuple is classified in the class with the highest posterior probability.

Major drawbacks of bayesian classification are the high computational complexity and the need for complete knowledge of prior and conditional probabilities.

3.3 Decision Trees

Decision trees (DT) are the most popular technique for prediction. In order to successfully apply a decision tree algorithm, a well-defined set of classes (categories) and a training set of pre-classified data are necessary. Decision tree quality is highly associated with the classification accuracy reached on the training dataset, as well as with the size of the tree. DT algorithms are two-phase processes [Han and Kamber, 2001; Quinlan, 1986]:

1) **Building phase:** The training data set is recursively partitioned until all the instances in a partition belong to the same class

2) **Pruning phase:** Nodes are pruned to prevent overfitting and to obtain a tree with higher accuracy

The primary features of a decision tree are:

- **Root node:** The dataset attribute that is selected as the base to built the tree upon

- **Internal node:** An attribute that resides somewhere in the inner part of the tree

- **Branch descending of a node:** One of the possible values for the attribute the branch initiates

- **Leaf:** One of the predefined classes

A DT example is illustrated in Figure 2.4. This particular tree attempts to predict whether a game of golf will be played, depending on weather conditions. The classes are two (YES: the game will be conducted and NO: the game will be postponed).

During the building phase, part of the initial dataset is used and one of the dataset attributes is selected to classify the tuples upon. If some dataset attributes are misclassified, exception rules are introduced and the process is repeated until all the attributes are correctly classified.

All DT algorithms apply a splitting criterion, which is a metric normally calculated for each attribute. The minimum (or maximum) value of this metric points to the attribute on which the dataset must be split. These criteria, along with the most representative algorithms of each category are:

- **ID3 and C4.5:** Information gain algorithms [Colin, 1996; Quinlan, 1992]

- **DBLearn:** An algorithm that exploits domain understanding to generate descriptions of predefined subsets [Koonce et al., 1997].

- **CLS:** An algorithm that examines the solution space of all possible decision trees to some fixed depth. It selects a tree that minimizes the computational cost of classifying a record [Hunt et al., 1966]

- **SLIQ, RAINFOREST and SPRINT:** Algorithms that select the attribute to split on, based on the GINI index. These algorithms manipulate successfully issues related to tree storage and memory shortage [Mehta et al., 1996; Ganti et al., 1999; Shafer et al., 1996].

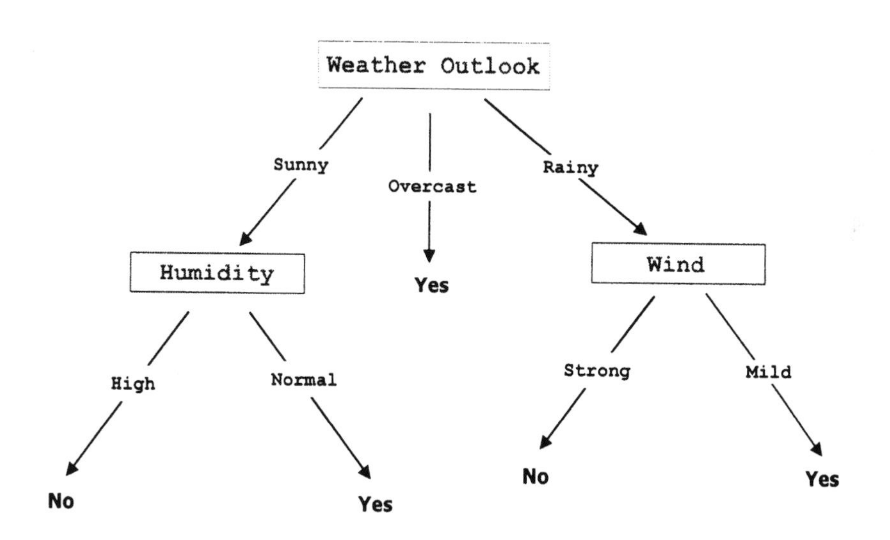

Figure 2.4. A sample decision tree

Decision trees are often used as inference engines or knowledge models to implement the behavior of agents. Since we are going to use them extensively in subsequent chapters, we present the ID3 algorithm in more detail, so that the unfamiliar reader can better comprehend the way DTs are built.

3.3.1 The ID3 algorithm

ID3 aims to minimize the number of iterations during classification of a new data tuple and can be applied only to categorical data. During the tree building phase, ID3 follows a top-down approach, which comprises the following three steps:

a) One of the dataset attributes is selected as the root node, and attribute values become tree branches.

b) All the tuples of the training dataset are placed on the branches with respect to the value of the root attribute. If all instances belong to the same class, then a node named by the class is created and the process terminates.

c) Otherwise, another attribute is selected to be a node, attribute values become tree branches and the process iterates.

Although all attributes are node candidates, a statistical property, the *information gain*, is used to choose the best one. The expected information gain when splitting dataset D with respect to attribute A_i, $A_i \in \mathbf{A}$ ($\mathbf{A}$ is the set of attributes describing D) is given by Eq. 2.5:

$$Gain(D, A_i) = Info(D) - Info(D, A_i) \tag{2.5}$$

$Info(D)$ is the information needed to classify D into the predefined distinct classes c_i (for $i = 1...N$), and is given by Eq. 2.6:

$$Info(D) = -\sum_{i=1}^{N} p(I) \log_2 p(I) \tag{2.6}$$

with $p(I)$ the ratio of D tuples that belong to class c_i.

$Info(D, A_i)$ is the information needed in order to classify D, after its partitioning into subsets D_j, $j = 1...\nu$, with respect to the attribute A_i. $Info(D, A_i)$, which is also denoted as the *Entropy of* A_i, is given by Eq. 2.7:

$$Info(D, A_i) = \sum_{j=1}^{\nu} \frac{|D_j|}{|D|} \times Info(D_j) \tag{2.7}$$

During the building phase, splitting is conducted on the attribute that produces the maximum information gain, while the attribute with the overall maximum information gain is selected as the root node.

Let us apply ID3 on the *Play Golf* example earlier mentioned. Table 2.2 summarizes the *Play Golf* dataset. In order to decide on the root node, we calculate the information gain for each dataset attribute (*Weather Outlook*, *Temperature*, *Humidity*, *Wind*):

Table 2.2. The *Play Golf* dataset

Weather Outlook	Temperature	Humidity	Wind	Play_ball
Sunny	High	High	Mild	No
Sunny	High	High	Strong	No
Overcast	High	High	Mild	Yes
Rainy	Normal	High	Mild	Yes
Rainy	Low	Normal	Mild	Yes
Rainy	Low	Normal	Strong	No
Overcast	Low	Normal	Strong	Yes
Sunny	Normal	High	Mild	No
Sunny	Low	Normal	Mild	Yes
Rainy	Normal	Normal	Mild	Yes
Sunny	Normal	Normal	Strong	Yes
Overcast	Normal	High	Strong	Yes
Overcast	High	Normal	Mild	Yes
Rainy	Normal	High	Strong	No

$$Gain(S,\ \textit{Weather Outlook}) = 0.246$$
$$Gain(S,\ \textit{Temperature}) = 0.029$$
$$Gain(S,\ \textit{Humidity}) = 0.151$$
$$Gain(S,\ \textit{Wind}) = 0.048$$

Since *Weather Outlook* yields the maximum information gain, it is selected as the root node. Its attribute values form the branches initiating from the root (Figure 2.5).

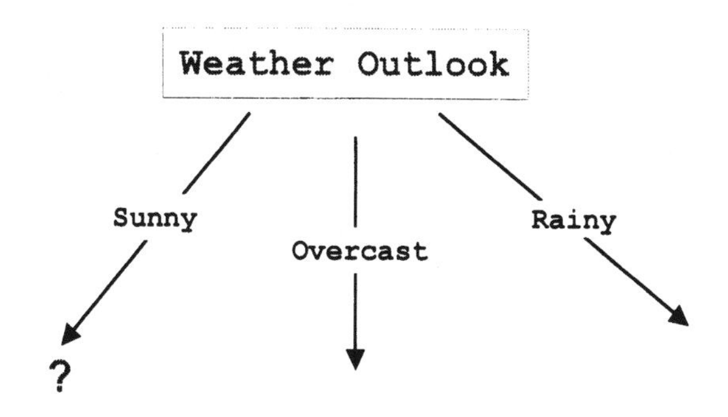

Figure 2.5. Deciding on the root node

For each one of the root branches and the remaining attributes the previous process is repeated, until the tree of Figure 2.4 is constructed.

The C4.5 algorithm extends ID3 by handling missing values and numerical data, by improving decision tree pruning, and by deriving classification rules.

Although quite popular, decision trees have a major scalability problem. Classification accuracy and execution time may be acceptable for small datasets, but, in the case of greater datasets, either execution time increases, or classification accuracy decreases substantially.

4. Clustering

Generally, clustering aims to divide a dataset into smaller "uniform" datasets, in order to better comprehend the data space, to simplify the analysis, and to enable the application of "divide-and-conquer"–type algorithms [Kaufman and Rousseeuw, 1990].

When it comes to data mining, clustering is performed on large datasets to identify groups or dense areas of data points, by the use of some distance metric. Data in the same clusters exhibit several common characteristics, while this is not the case for data in different clusters. In other words, *clustering is the process of identifying a finite set of categories or clusters to describe the data.* Figure 2.6 illustrates the result of clustering on a small dataset.

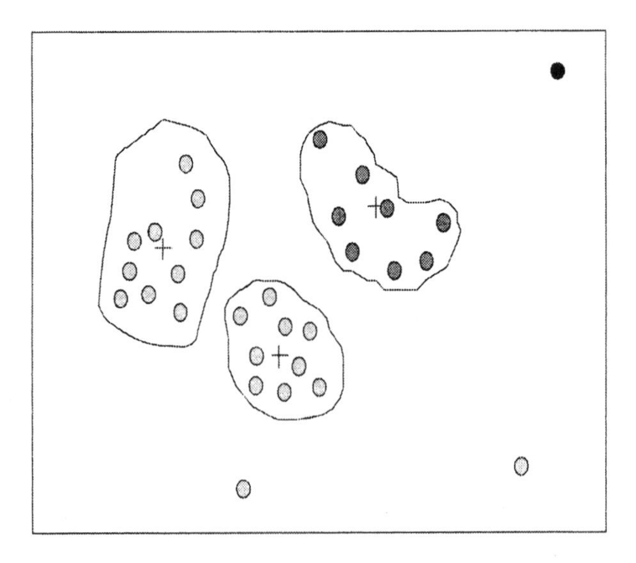

Figure 2.6. The clustering concept. The dataset is divided into three multiple-data point and three single-data point clusters.

4.1 Definitions

We define a *cluster* as the collection of data objects that are similar to one another within the same cluster *and* dissimilar to the objects in other clusters.

In contrast to classification, which is a supervised learning technique, clustering is unsupervised. There are no predefined classes and no examples to indicate possible relations among the data.

In order for clustering to be successful, the following conditions must be satisfied (Figure 2.7) [Han and Kamber, 2001]:

- High *intra-cluster* similarity
- Low *inter-cluster* similarity

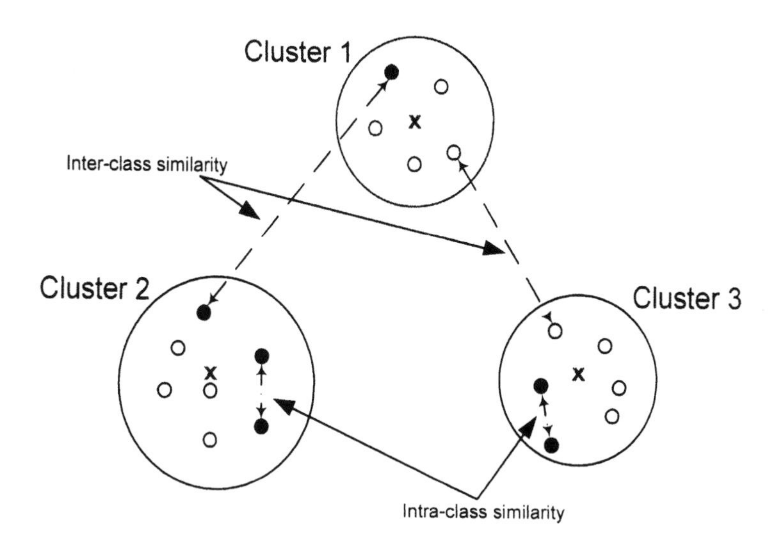

Figure 2.7. Intra- and inter-cluster similarity

Additional requirements for successful clustering include algorithm scalability and reusability, multi-dimensional data manipulation, discovery of clusters with arbitrary shapes, ability to deal with noise and outliers, and ability to process various data types.

4.2 Clustering Techniques

Existing techniques for clustering can be divided into categories using three criteria [Jain et al., 1999]:

a) Based on the method used to identify clusters, they can be classified into partitional, hierarchical, density-based, and grid-based algorithms.

b) Based on the type of data they manipulate, they can be classified into statistical and conceptual algorithms.

c) Based on the theory used to extract clusters, we can have fuzzy clustering, crisp clustering and Kohonen-net clustering algorithms.

In the following section we briefly describe the most representative partitional, hierarchical, and density-based algorithms.

4.3 Representative Clustering Algorithms

4.3.1 Partitioning Algorithms

Partitioning algorithms decompose the dataset into disjoint clusters. The basic problem that this type of clustering algorithms solves is the following:

> Given an integer k, find a partition of k clusters $c_1, c_2, ..., c_k$ that optimizes the chosen partitioning criterion.

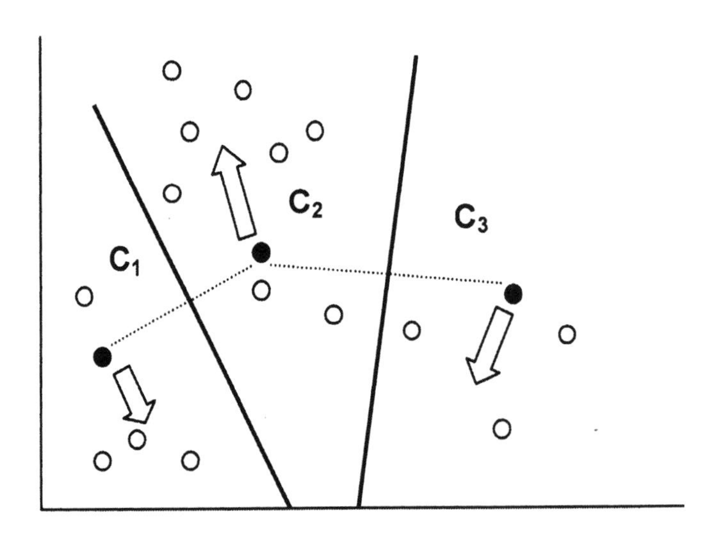

Figure 2.8. K–Means schematic representation

The most representative partitioning algorithm is the *K–Means* algorithm [Adriaans and Zantinge, 1996; McQueen, 1967]. The *K–Means* mechanism is quite straightforward: First, k data objects are randomly selected as the centers of the k clusters. The rest of the data objects are assigned to the closest one of the already formed clusters. Next, for each cluster c_i, $i = 1...k$, the median, m_i, is calculated and becomes the new cluster center. This process is repeated until a predefined selection function converges. A typical selection function is the square error, defined as:

$$E = \sum_{i=1}^{k} \sum_{p \in c_i} |p - m_i|^2 \tag{2.8}$$

where E is the total square error for all data objects p of the dataset. Figure 2.8 provides a schematic representation of the algorithm. The space is divided into three clusters C_1, C_2, C_3 with the cluster centers standing out.

The main advantages of K–Means include termination at a local optimum and a relatively good efficiency, since algorithm complexity is $O(nkt)$, where n is the number of data objects, k is the number of clusters, and t is the number of iterations till convergence.

A major limitation of K–Means is that it can be applied only to numerical datasets. Other drawbacks include the need to specify the number of clusters in advance, the algorithm's poor handling of noisy data and outliers, and its inability to discover clusters with non-convex shapes.

To overcome some or all of the K–Means drawbacks, several other algorithms have been developed. Instead of using cluster medians, one can use cluster *medoids*. A medoid is defined as the most central data object within a cluster. *CLARANS* [Chen et al., 1996; Ng and Han, 1994] is a typical medoid algorithm, based on the *PAM* and *CLARA* algorithms [Kaufman and Rousseeuw, 1990]. It uses random sampling to identify the optimal k clusters. Once the k medoids are specified, the remaining data points are assigned to the cluster corresponding to their closest medoid. Next, the possibility of replacing a medoid with any of its neighbors is examined. If some other data point improves the overall cluster quality, it becomes the new medoid and the process is repeated. If not, a local optimum is produced and CLARANS randomly selects other data objects in its quest to find an overall optimum. It should be mentioned that CLARANS' complexity increases linearly with the number of data points.

Based on CLARANS, two more algorithms that can manipulate spatial data have been introduced. For algorithms *SD(CLARANS)* and *NSD(CLARANS)* [Ng and Han, 1994], the user has to ability to specify the type of rules to be extracted, through a learning process.

4.3.2 Hierarchical Algorithms

Hierarchical algorithms construct trees where each node represents a cluster. They can be further discriminated into cumulative and divisive algorithms, following a bottom-up or a top-down approach, respectively. The most popular algorithms of this category are:

- **Birch:** An algorithm that uses CF-trees (defined below) and incrementally adjusts the quality of sub-clusters [Zhang et al., 1996].
- **CURE:** A robust to outliers algorithm that can identify clusters of non-spherical shapes [Guha et al., 1998].
- **ROCK:** A robust clustering algorithm for Boolean and categorical data. It introduces two new concepts: a point's *neighbors* and its *links* [Guha et al., 2000].

Since Birch is the most representative hierarchical algorithm, we provide a brief description of it.

Birch (Balanced Iterative Reducing and Clustering Hierarchy) is used to cluster large datasets. Two are its main characteristics: a) the clustering feature *CF* and b) the *CF–tree.*

Let there be N d-dimensional data tuples or data objects o_i in a subcluster. CF summarizes information about subclusters of objects and is defined as a triplet $(N, \overrightarrow{LS}, SS)$, where N is the number of data objects residing within the cluster, $\overrightarrow{LS}$ is the sum of the data objects $(\sum_{i=1}^{N} \vec{o_i})$ and SS is their square sum $(\sum_{i=1}^{N} \vec{o_i}^2)$.

The CF–tree is a weighted tree with two parameters: the branching factor B and threshold T. B determines the maximum number of children a tree node may have, while T determines the maximum diameter of a subcluster residing in the leaves. The CF–tree is built incrementally, while data are provided as input. A new data object is assigned to its closest subcluster. If the subcluster size exceeds T, then the node is split into two. The information of a new data object insertion, percolates up until the root of the tree. By modifying T the tree size can be adjusted. In that case the tree is not reconstructed, since new leaves are created from the old ones. CF–trees can be used to reduce the complexity of any algorithm. In general, Birch scales linearly with the number of data objects, is independent of the data input sequence, and can usually produce good clustering results with a single scan, improving its quality with a few additional scans.

4.3.3 Density-Based Algorithms

Clustering can also be performed with respect to the density of the data objects within each cluster. Density-based algorithms have the ability to detect clusters of arbitrary shape, they can handle noise, while they need a density-based condition to terminate. Representative algorithms in this category are *DBSCAN* [Ester et al., 1996] and *DENCLUE* [Hinneburg and Keim, 1998]. A brief description of the former follows.

Within the context of DBSCAN, the cluster is defined as the maximal set of density-connected points. DBSCAN succeeds greatly in identifying clusters of arbitrary shape in spatial databases with noise. Density-based clustering entails the definition of new concepts:

- Given a data object p, data objects that are within an *Eps* radius are called *Eps-neighborhood* of p.
- If the *Eps-neighborhood* of p contains at least *MinPts* data points, then p is denoted as a *core point.*
- Given a dataset D, we define p to be *directly density-reachable* from data point q, if p lies within the *Eps-neighborhood* of q and q is a core point.
- Data point p is *density-reachable* from q (with respect to *Eps* and *MinPts*) within dataset D, if a sequence $p_1, ..., p_n$ $(p_1 = q, p_n = p)$ of data points exists, so that $p_i + 1$ is *directly density-reachable* from p_i (with respect to *Eps* and *MinPts*), for $1 \leq i \leq n$, $p_i \in D$.
- Data point p is *density-connected* to q with respect to *Eps* and *MinPts* in D, if there is one data point in D for which p and q are *density-reachable.*
- A *density-based cluster* is the maximum set of density-connected data points. If a data point is not assigned to any cluster, it is characterized as an *outlier.*

Figure 2.9 serves as an example of the above definitions. The DBSCAN mechanism can be summarized in the following steps [Han and Kamber, 2001]:

1. Randomly select a data point p
2. Retrieve all the *density-reachable* points, with respect to *Eps* and *MinPts*
3. If p is a core point, a cluster is formed
4. For each core point, find all *density-reachable* points
5. Merge all *density-reachable* clusters
6. If p is an outlier, go to the next data point
7. Repeat until no other data point can be inserted to any cluster.

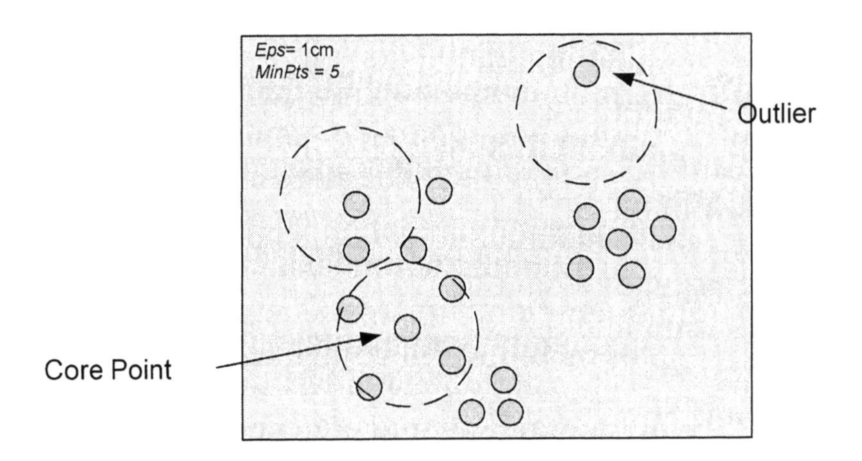

Figure 2.9. The concepts of DBSCAN

5. Association Rule Extraction

In transactional databases, it is often desirable to extract rules which assert that occurrence of subset I_A within a transaction implies the occurrence of subset I_B within the same transaction. Rules of this form are typical in association rule extraction.

> *We define* ***association rule extraction*** *as the process of finding frequent patterns, associations, correlations, or causal structures among sets of items or objects in transaction databases, relational databases, and other information repositories* [Amir et al., 1999].

5.1 Definitions

Let $I = i_1, i_2, ...i_m$ be a set of data objects called *items*. Let D be the transaction set, where each transaction T is a collection of items,with $T \subseteq I$. Let I_A be a set of items. We say that transaction T contains I_A, iff $I_A \subseteq T$. An *association rule* is a relationship of the form $I_A \Rightarrow I_B$, where $I_A \subseteq I$, $I_B \subseteq I$ and $I_A \cap I_B = \varnothing$. This rule is considered to have *support* s, where s is the transaction percentage in D that contain $I_A \cup I_B$ (i.e., either I_A or I_B). Consequently, s is the probability $P(I_A \cup I_B)$. Rule $I_A \Rightarrow I_B$ also has *confidence* c within D, if c is the percentage of transactions in D that contain both I_A and I_B. That is, c is the conditional probability $P(I_A|I_B)$. Summarizing, *support* and *confidence* are given by the following equations [Zhang and Zhang, 2002; Agrawal and Srikant, 1994]:

Table 2.3. A sample transaction database

Transaction ID	**Items Bought**
2000	A,B,C
1000	A,C
4000	A,D
5000	B,E,F

$$support(I_A \Longrightarrow I_B) = P(I_A \cup I_B) \tag{2.9}$$

$$confidence(I_A \Longrightarrow I_B) = P(I_A | I_B) \tag{2.10}$$

Rules that satisfy the $s \geq min\_sup$ and $c \geq min\_conf$ conditions, with min_sup and min_conf given thresholds, are considered *strong*.

A set of items is called an *itemset*. Thus, a set that contains k items is called a $k - itemset$. A $k - itemset$ is defined as *frequent*, when for its items the $s \geq min\,\sup$ relation holds. It should be noted that all the subsets of the $k - itemset$ are also frequent.

The following example attempts to clarify the concepts of *support* and $confidence$.

Let there be the transactional dataset illustrated in Table 2.3. We would like to calculate the s and c for the rules $A \Rightarrow C$ and $C \Rightarrow A$.

We observe that both A and C appear in two out of four transactions. Support s is therefore $s = 2/4 = 50\%$.

As far as the first rule is concerned ($A \Rightarrow C$), we observe that A appears in three transactions, while C also appears into two of them. Thus, confidence $c = 2/3 = 66\%$.

For the second rule ($C \Rightarrow A$), C appears in two transactions and in both of these transactions A also appears. Thus, confidence $c = 2/2 = 100\%$.

5.2 Representative Association Rule Extraction Algorithms

The most representative association rule extraction algorithms are [Han and Kamber, 2001]: a) Apriori, b) DHP, c) Trie Data Structures, and d) Iceberg queries.

The Apriori algorithm is presented next, to provide some insight into the way association rules are extracted.

Apriori algorithm

As the name of the algorithm implies, a priori knowledge on the already formed frequent itemsets is used. Apriori [Agrawal and Srikant, 1994] implements an iterative process where $k - itemsets$ are used, in order to discover $(k+1) - itemsets$. First, all frequent $1 - itemsets$, L_1, are calculated. L_1 is then used in order to discover L_2, the set of frequent $2 - itemsets$. L_2 is then used to calculate L_3, and so forth, until no higher rank itemsets can be discovered.

Apriori employs a two–step process:

1) **Joining:** In order to calculate L_k, a set of candidate $k - itemsets$ is created, by joining L_{k-1} with itself. This set is denoted as C_k.

2) **Pruning:** C_k is the hyperset of L_k, containing both frequent and non-frequent itemsets. Only the frequent $k - itemsets$ form L_k. The process iterates until no higher rank itemsets can be discovered.

Figure 2.10 illustrates a schematic representation of the Apriori process.

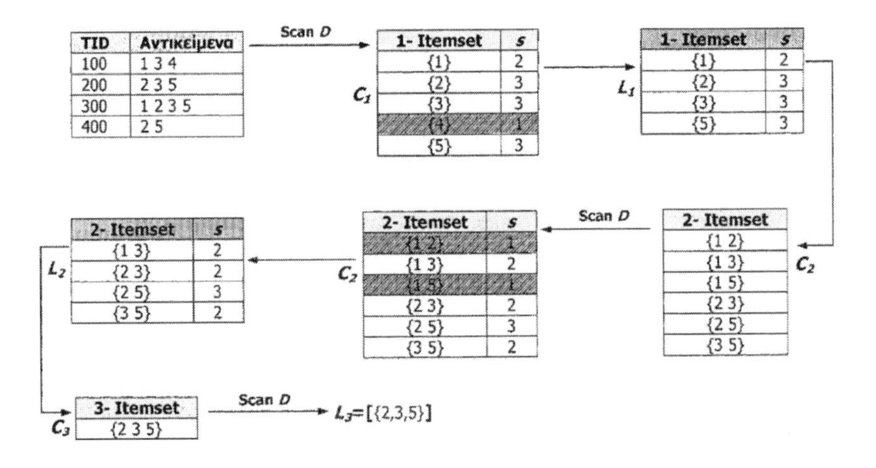

Figure 2.10. A schematic representation of the Apriori algorithm. Only itemsets with $s \geq 2$ are retained for the next iteration.

Although effective, Apriori presents a scalability problem and it cannot handle numerical data. These issues are dealt by other algorithms that extend Apriori and are briefly mentioned below:

- **DHP:** The algorithm creates a hash table that controls the legitimacy of $k - itemsets$. Instead of using the L_{k-1} joined by itself to produce C_k, DHP includes only the $k - itemsets$ in the hash table that have support $s \geq min\_sup$. This way both the size of C_k and the database size are reduced dramatically [Chen et al., 1996].

- **FTDA:** It is used in order to extract association rules on quantitative values. Fuzzy methods are applied and fuzzy rules are extracted [Hong et al., 1999].

- **Partition:** The algorithm divides the itemset database into two segments and requires only two scans. During the first scan, Partition discovers the frequent itemsets for each one of the two segments, while it calculates the support of each itemset with respect to all itemsets [Ganti et al., 1999].

- **DIC:** It optimizes association rule extraction through dynamic itemset counting [Ganti et al., 1999].

- **Trie Data Structure:** Used to create *covers*, i.e., itemsets that have greater than or equal to a specified minimum support threshold. The most interesting feature of these structures is their ability to extract *exclusive* association rules [Amir et al., 1999].

6. Evolutionary Data Mining Algorithms

6.1 The Basic Concepts of Genetic Algorithms

Genetic Algorithms (GAs) aim to either maximize or minimize a function and can, therefore, be classified as optimization algorithms. GA technology goes back to the 1950's, when a group of biologists attempted to use computers in order to simulate ecosystems. Early research on GAs was performed by John Holland and his students at the University of Michigan, during the 60's and 70's.

Research on species evolution has indicated that living organisms have an inherent ability to adapt dynamically to a constantly varying environment. This fact led Holland to simulate this process through an algorithm called a *Genetic Algorithm* [Holland, 1975; Holland, 1987; Booker et al., 1989]. GAs attempt to mimic the robustness, efficiency and flexibility of biological systems. Along with self-repair and reproduction, which are core functionalities in nature, GAs bypass the problems often encountered by artificial systems.

Traditional optimization techniques either search for local optima, or focus on the calculation of the function derivatives. Such techniques cannot deal efficiently with more complex problems, in contrast to GAs. The basic differences of GAs from the rest of the optimization techniques are [Mitchell, 1996]:

1. GAs process an encoded form of the parameters of the problem at hand and not the parameters themselves. Binary encoding is usually applied on a finite length string.

2. GAs search the solution among a set of candidates and not by targeting a specific one. In most optimization techniques, we shift from one point of the solution space to another following some transition rules. Such techniques cannot be applied to complex search spaces that are not continuous and non-linear, since they often result to local optima. The deployment of a set of solution candidates reduces the probability of arriving at such a local optima.

3. GAs follow probabilistic rules and random selection in order to search the space where the optimal values lie.

4. GAs do not require prior knowledge of the search space, in contrast to other approaches. Gradient techniques, for example, require the existence and continuity of derivatives. GAs solely rely on their *fitness function.*

The above variations of GAs with respect to traditional techniques provides them with the desired robustness.

6.2 Genetic Algorithm Terminology

All the terms of GAs originate from the fields of biology and genetics [Goldberg, 1989]. For example, as already mentioned, problem parameters are encoded as a *bit* sequence. This bit sequence that comprises the *string* is analogous to the *chromosomes* in biology. In nature, two or more chromosomes are combined to finally create the *genotype.* In GAs, genotypes are mapped to the *structure,* the set of available strings. In nature, the organism developed when the genetic code interacts with the surrounding environment, is called a *phenotype,* which in the case of GAs corresponds to the decoded structure of the problem, aiming to create a candidate solution space. In biological systems chromosomes comprise *genes,* which are responsible for one or more organism characteristics. A gene can fall into a number of different states, called *alleles.* In the case of genetic algorithms, genes are mapped to bits, with their alleles being 0s and 1s. Finally, gene position within a chromosome is called *locus,* while bit position is called *position.* Table 2.4 lists the main features of GAs and their mapping to genetics terms.

Table 2.4. The core features of Genetic Algorithms

Nature	Genetic Algorithm
Chromosome	String
Gene	Bit
Allele	Bit value
Locus	Bit position within String
Genotype	Structure (Set of Strings)
Phenotype	Decoded Structure
Being	Solution

6.3 Genetic Algorithm Operands

The genetics-related operations performed by living organisms are mapped to three GA operands, namely *selection*, *crossover*, and *mutation.*

The *selection* operand is responsible for the survival of the fittest solution. Based on the fitness function of the GA, strong chromosomes have a higher probability to contribute one or more genes to the upcoming generations.

The *crossover* operand combines genes from "capable" chromosomes, in order to produce even fitter ones. Crossover follows selection and is a three-step process (Figure 2.11):

1. Recently reproduced strings are randomly selected in pairs.
2. A random position is selected for each pair
3. Based on some crossover probability, the pairs exchange part of their bit sequences on the position previously selected.

Finally, the mutation operand mimics the genetic mutation that is responsible for random changes in the genetic code. Just like in biological systems, mutation plays a substantial, nevertheless small part in the evolution of populations. The operand modifies the state of one or more genes, based on a mutation probability. Figure 2.12 illustrates the result of mutation for a chromosome at a randomly selected position.

Mutation, working in close collaboration with the operations of selection and crossover, improves the algorithm's ability to identify new patterns. It prevents the GAs from terminating on local optima, by inserting new, not yet explored chromosomes in the population.

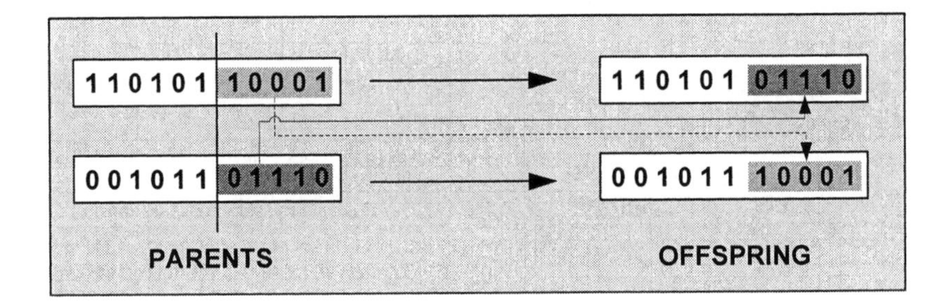

Figure 2.11. Chromosome crossover

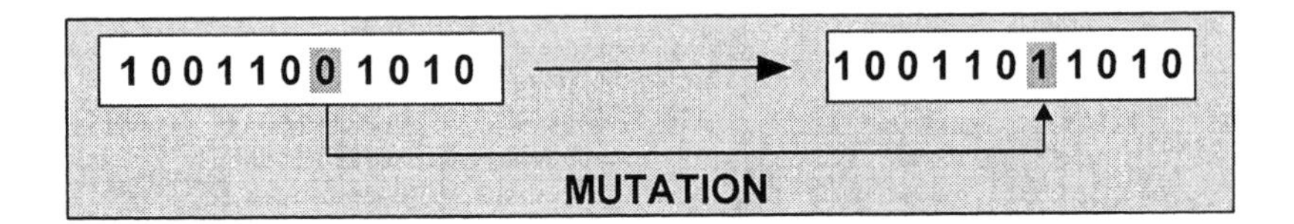

Figure 2.12. Chromosome mutation

6.4 The Genetic Algorithm Mechanism

Figure 2.13 provides a schematic representation of the GA mechanism. First, the algorithm instantiates a chromosome population. In order to give birth to the next generation, the reproduction, crossover and mutation operands are applied. For each new chromosome, a value for the fitness function is computed, which shall in turn specify whether the chromosome will be selected for reproduction. This process iterates until an optimal solution is found or until a predefined terminating condition is satisfied.

6.5 Application of Genetic Algorithms

Genetic algorithms have been used in a wide range of applications. The most characteristic ones are outlined below [Mitchell, 1996]:

- **Optimization:** GAs have been employed for optimization problems, including numerical optimizations, as well as combinatorial problems.

- **Automated Programming:** GAs have been used for the improvement of developed software, as well as for the design of specific computational structures (cellular automata, sorting networks).

- **Machine Learning:** GAs have been widely adopted in machine learning applications, including classification and prediction problems (weather forecasting, protein structure classification). They have also

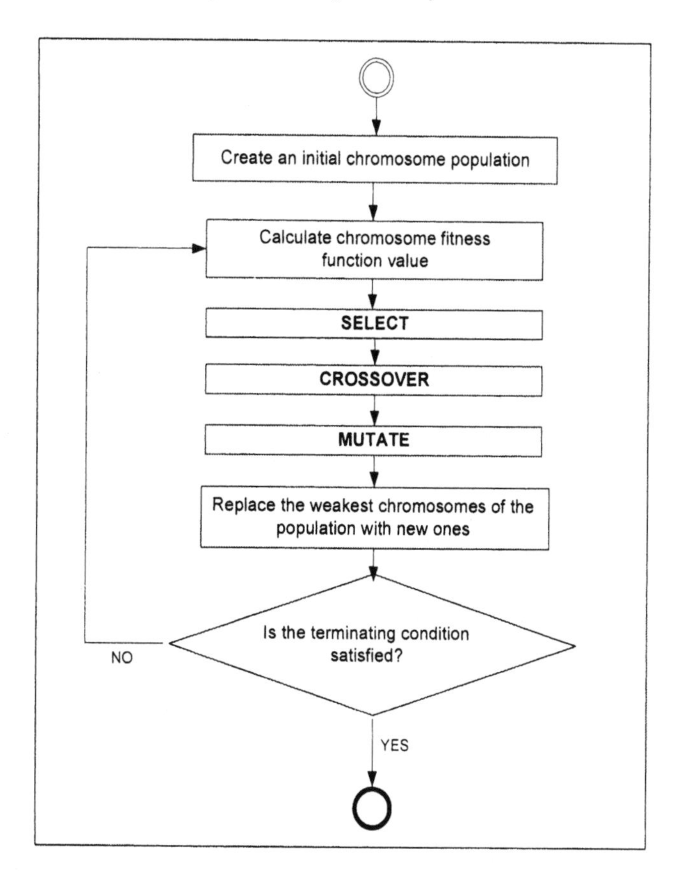

Figure 2.13. The genetic algorithm mechanism

been used for the optimization of machine learning algorithm parameters, as in the case of neural network weights.

- **Economics:** Pricing policies, financial market simulations, and risk assessment have been performed with the use of GAs.
- **Ecology:** GAs have been employed for the modeling and simulation of ecological problems, like species survival, symbiosis etc.
- **Population genetics:** The study of populations and their evolution has been supported by the use of GAs.
- **Evolution and learning:** GAs have been used to improve the interaction between the members of a population.
- **Social Modeling:** Finally, GAs have been used for the study of societal issues and their impact on populations.

Within the context of the book, genetic algorithms are employed in order to either extract or improve knowledge in learning systems, following the Michigan approach [Goldberg, 1989], where each member of the GA population represents a rule, i.e., part of the solution.

7. Chapter review

The aim of this Chapter was to introduce the reader with the data mining concepts and techniques this book deals with. Useful definitions are provided, in order to avoid any misunderstanding, while an overview of the most characteristic DM preprocessing methods is provided. The principles of classification, clustering, association rule extraction and genetic algorithms, along with their most representative algorithms are presented.

Chapter 3

INTELLIGENT AGENTS AND MULTI-AGENT SYSTEMS

1. Intelligent Agents

1.1 Agent Definition

Although the word agent is used in a multitude of contexts and is part of our everyday vocabulary, there is no single all-encompassing meaning for it. Perhaps, the most widely accepted definitions for this term is that "*an agent acts on behalf of someone else, after having been authorized*". This definition can be applied to *software agents*, which are instantiated and act instead of a user or a software program that controls them. Thus, one of the most characteristic agent features is its *agency*. In the rest of this book, the term agent is synonymous to a software agent.

The difficulty in defining an agent arises from the fact that the various aspects of agency are weighted differently, with respect to the application domain at hand. Although, for example, *agent learning* is considered of pivotal importance in certain applications, for others it may be considered not only trivial, but even undesirable. Consequently, a number of definitions could be provided with respect to agent objectives.

Wooldridge & Jennings have succeeded in combining general agent features into the following generic, nevertheless abstract, definition:

> *An agent is a computer system that is situated in some environment, and that is capable of autonomous action in this environment, in order to meet its design objectives* [Wooldridge and Jennings, 1995].

It should be denoted that Wooldridge & Jennings defined the notion of an "*agent*" and not an "*intelligent agent*". When intelligence is introduced, things get more complicated, since a nontrivial part of the research community believes that true intelligence is not feasible [Nwana, 1995]. Nevertheless, "*intelligence*" refers to computational intelligence and should not be confused with human intelligence [Knapik and Johnson, 1998].

A fundamental agent feature is the *degree of autonomy*. Moreover, agents should have the ability to *communicate* with other entities (either similar or dissimilar agents) and should be able to exchange information, in order to reach their goal. Thus, an agent is defined by its *interactivity*, which can be expressed either as proactivity (the absence of passiveness) or reactivity (the absence of deliberation) in its behavior. Finally, agents can be defined through other key features, such as their learning ability, their cooperativeness, and their mobility [Agent Working Group, 2000].

One can easily argue that a generic agent definition, entailing the above characteristics, cannot satisfy researchers in the fields of Artificial Intelligence and Software Engineering, since the same features could be ascribed to a wide range of entities (e.g., man, machines, computational systems, etc.). Within the context of this book, agents are used as an abstraction tool, or a metaphor, for the design and construction of systems [Luck et al., 2003]. Thus, ***no distinction between agents, software agents and intelligent agents is made***. A fully functional definition is employed, integrating all the characteristics into the notion of an agent:

> *An agent is an autonomous software entity that – functioning continuously – carries out a set of goal-oriented tasks on behalf of another entity, either human or software system. This software entity is able to perceive its environment through sensors and act upon it through effectors, and in doing so, employ some knowledge or representation of the user's preferences* [Wooldridge, 1999].

1.2 Agent Features and Working Definitions

Bypassing the problem of proper definitions, one can focus on the functional characteristics of agents. An agent may exhibit, depending on the problem it is confronted, with some or all of the properties listed below [Wooldridge and Jennings, 1995; Genesereth and Ketchpel, 1994; Agent Working Group, 2000]:

a. **Autonomy:** An agent may act without constant surveillance and direct human (or other) intervention and may have some kind of control over its actions and internal states, based on its beliefs, desires, and intentions. Autonomy is considered a prerequisite by many researchers.

b. **Interactivity:** An agent may interact with its environment and other entities. Interactivity entails all the behavioral aspects that an agent may exhibit within an environment. Two are the most dominant features that univocally determine an agent behavior's state:

   i) **Reactivity:** The ability to perceive the environment and its changes, and to respond to them, when necessary.

   ii) **Pro-activeness:** The ability to take initiative and act in such a way to satisfy the goal the agent has been deployed for.

c. **Adaptability:** An agent may perceive the existence of other agents within its "sight" or "neighborhood". Advanced adaptability schemes allow agents to alter their internal states based on their experience and their environment.

d. **Sociability:** It is of great importance for agents to develop a behavior model that resembles aspects of human social life, such as companionship, friendship and affability. Agent interaction is established via some kind of communication language.

e. **Cooperativity:** The ability to collaborate in order to accomplish a common goal. This mode of operation is highly desirable in communities of agents seeking the same "*holy grail*".

f. **Competitiveness:** An agent may compete with another agent, or deny service provision to an agent perceived as an opponent. This behavior is required in situations where only one agent can reach the target (i.e., electronic auctions).

g. **Temporal continuity:** Since an agent functions continuously, proper design to ensure robustness is essential.

h. **Character:** An agent may exhibit human–centered characteristics, such as personality and emotional state.

i. **Mobility:** The ability to transit from one environment to another, without interrupting its functioning.

j. **Learning:** An agent should be able to learn (get trained) through its reactions and its interaction with the environment. This is one of the most fundamental features that agents, presented within the context of this book, should have. The more efficient the training process, the more intelligent the agents.

Agents could also be defined with respect to the domain in which they provide their services [Knapik and Johnson, 1998]. Typical domains of agent use are listed below:

- Searching for information
- Filtering data
- Monitoring conditions and alerting users on the detection of set point conditions
- Performing actions on behalf of a user, i.e., personal agents
- Providing data access and transactional security
- Providing context-sensitive help
- Coordinating network tasks and managing resources
- Optimizing system use via goal-oriented techniques

1.3 Agent Classification

By now it should be evident that agents can be classified in a variety of ways, taking into account the subset and the importance of features they exhibit.

According to Nwana [Nwana, 1995], agents can be classified with respect to the following dimensions:

i. *Mobility*, that differentiates agents into *static* or *mobile*

ii. The *logic paradigm* they employ, which classifies them as either *deliberative* or *reactive*

iii. The *fundamental characteristic* that describes the agent (autonomy, cooperativity, learning). Based on these axes, agents can be classified as *collaborative agents*, *collaborative learning agents*, *interface agents* and *truly smart agents* (Figure 3.1).

The combination of two or more of the above approaches classifies agents as *hybrid*, leading to *mobile deliberative collaborative agents*, *static reactive collaborative agents*, *static deliberative interface agents*, *mobile reactive interface agents*, etc.

Finally, agents can be classified upon *secondary characteristics*, i.e., competitiveness, trustworthiness, benevolence etc.

Although fuzzy and versatile, Nwana classification yields seven distinct agent classes: *collaborative agents*, *interface agents*, *mobile agents*, *information/internet agents*, *reactive agents*, *hybrid agents* and *smart*

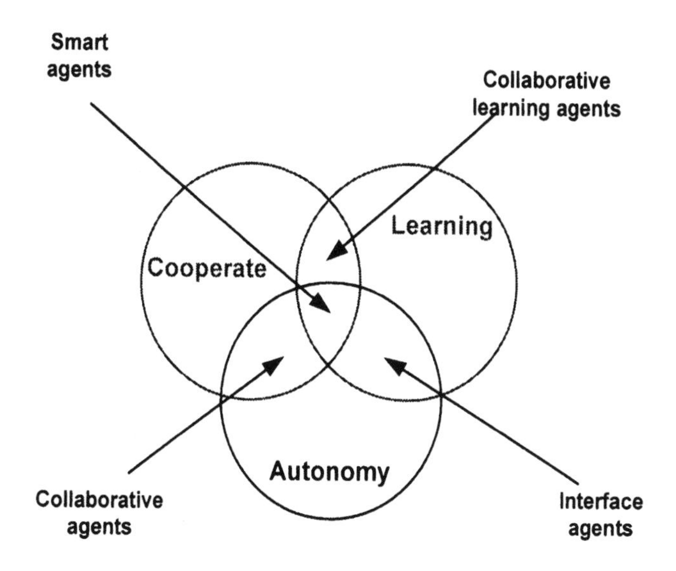

Figure 3.1. The Nwana agent classification, according to their fundamental characteristics

agents. Applications that deploy agents from more than one of these classes are denoted as *heterogeneous agent systems.*

Nwana's classification scheme is robust enough to meet the needs and demands of a large number of IA researchers, while it also covers a wide area of agent-related applications. Therefore, we have also adopted it and use it throughout the book.

1.4 Agents and Objects

One of the most common pitfalls for agent technology is the confusion between agents and objects. An object (within the context of object-oriented programming) is often defined as a computational entity that encapsulates some state and is able to perform actions or methods on this state in order to alter it. Moreover, an object has the ability to communicate to other objects via message passing [Wooldridge, 1999].

This definition illustrates a number of similarities between agents and objects, leading the reader in confusion. Obviously, the relationship between an agent and an object depends on the multiple angles the two entities can be viewed. The Agent Working Group, for example, discusses both aspects (agents differ from objects, agents are objects themselves) [Agent Working Group, 2000], without providing a straight answer.

Although similar in many respects, agents differ from objects in three major aspects [Wooldridge, 1999]:

a) Agents embody a stronger notion of autonomy than objects. They have the ability to decide on the execution of an action, upon request by another agent. Objects do not have freedom of choice.

b) Agents are capable of flexible (reactive, pro-active, social) behavior, in contrast to objects.

c) A multi-agent system is inherently parallel and multi-threaded, having multiple loci of control (each agent has at least one thread of control).

It is common practice for researchers to promote the use of object-oriented technologies for the development of agents, called the OO–agents [Tschudin, 1995]. Such an approach offers many advantages, with most important being the following [Knapik and Johnson, 1998]:

1. **Reusability**
   Agents that comprise objects can be reusable, either as part or as a whole. Reusable agents can be utilized in more generic applications and can be employed to solve more than one problems, thus reducing implementation costs.

2. **Flexibility**
   Agents become very flexible by the use of objects, since the latter provide modularity. OO–agents can be deployed in a more or less functional mode, depending on the application and user's needs.

3. **Maintainability**
   An object-oriented approach of building agents leads to more maintainable architectures, especially in the case of large-scale systems.

4. **Extensibility**
   This approach enables extensible agent system architectures, allowing the creation and development of new agents types, either by the original software designer, or the end–user. Embedded inheritance, modularity and polymorphism of objects simplify the processes of agent modification, enhancement, or even deletion with respect to traditional (structured) development approaches.

5. **Reduction of Agent-developing costs**
   All of the above advantages of object-oriented programming in developing agents, lead to the reduction of the overall design, development and implementation costs.

Agents developed within the context of the book, follow this object-oriented approach.

1.5 Agents and Expert Systems

Expert systems, which constituted the mainstream AI technology of the 80's [Hayes-Roth et al., 1983], are computer applications developed to carry out tasks that would otherwise be performed by a human expert [Quinlan, 1987]. They are capable of solving problems in some knowledge–rich domain [Gondran, 1986] and the expertise of this domain is stored in data repositories that compose a *knowledge base.* This knowledge base is usually augmented by a set of rules, which are fired sequentially (based on a predefined scheme), to produce the system output in the form of recommendations. Expert systems often incorporate a mechanism for creating new rules in order to efficiently handle complex questions or scenarios. Some expert systems are designed to take the place of human experts, while others are designed to assist them in their decisions.

Expert systems could be considered the predecessors of agent systems and, thus, the two technologies are quite similar. The most important distinction between agents and expert systems is that the latter are inherently disembodied from their environment, not interacting directly with it [Wooldridge, 1999]. Rather, they provide feedback to third parties. Additionally, unlike agents, in expert systems cooperation is not a prerequisite. Representative expert systems include MYCIN [Shortliffe, 1976] and ARCHON [Jennings et al., 1996].

The use of existing expert systems can facilitate agent development. Direct access to pre-constructed data storehouses can be provided to an agent to obtain knowledge (and "intelligence") about a wide variety of subjects within a certain domain. In fact, the majority of existing expert systems' knowledge bases would be too large to be encapsulated within a single agent. However, if the amount of expertise needed is very focused, then a small set of rules could be programmed and embedded into a single or many agents.

1.6 Agent Programming Languages

Agent programming languages are evaluated with respect to the degree they facilitate the development of agents as software entities. The critical issue is how a fundamental characteristic of an agent can be depicted during its implementation. Since we have accepted that the object–oriented approach is more advantageous than the traditional programming paradigm, we focus on the corresponding languages that lend themselves to the construction of multi-agent systems.

Some of the languages mentioned below have limited use, since they did not receive sufficient support by large commercial projects. However, the choice of a programming language still remains an open issue for developers and is tightly dependent on the special needs of the agent designer.

Interesting efforts include Obliq [Cardelli, 1995], Aglets [Chang and Lange, 1996], Telescript [White, 1994], Smalltalk [Goldberg and Robson, 1983], and C++. These languages facilitated the implementation of agents in the early years of agent-based projects development (10-15 years ago).

The revolution of Java, released in 1995, affected dramatically agent software development. The main reason for Java's acceptance by agent developers is that it provides a set of capabilities that serve well the domain of distributed computing. Most important of these is the ability of Java to move and execute its byte-code to remote computers. Additionally, many APIs (Application Programming Interfaces) have been implemented to facilitate the development of agents, and many development tools for agents are available in Java. Today Java is considered as the "de-facto" programming tool for building agent-based applications.

2. Multi-Agent Systems

The emergence of distributed software architectures, along with the flourishing of the Grid computing [Fox et al., 2002; Roure et al., 2003] and Web Service initiatives [Sycara et al., 2003; Wang et al., 2004] have led to the development of Distributed Computational (DC) systems of the "new era". Advancing from naive architectures, where various processes collaborate to solve problems while competing for the available computational resources, DC systems are now open and heterogeneous, and their complexity is defined through the dynamics of interaction and reconfiguration [Frederiksson and Gustavsson, 2001]. Openness and heterogeneity make distributed computation a challenging task, since the processes involved must obtain resources in a dynamically changing environment and must be designed to collaborate, despite a variety of asynchronous and unpredictable changes. In the absence of global knowledge for determining the optimal resource allocation, approaches markedly different from ordinary system-level programming must be exploited [Huberman and Hogg, 1988]. Additional features of DC systems include a modular software architecture to ensure system scalability, the design of an incentive mechanism to entice third parties to modify the capabilities of the system, in order to ensure adaptability, coordination between the elements of the system, and a common communication structure [Birmingham, 2004]. Finally, special attention should also be drawn to the fact that in modern distributed computing, the inherent dynamism of networks makes self-awareness necessary [Guessoum,

2004]. The autonomous components of the system must be able to self-organize their activities' patterns to achieve goals, that possibly exceed their capabilities as single units, despite – and possibly taking advantage – of environment dynamics and unexpected situations [Parunak et al., 2001; Serugendo and Romanovsky, 2002].

All these key requirements are satisfied by the flourishing of the Agent-Oriented Software Engineering (AOSE) paradigm, which has provided the essential set of high-level and flexible abstractions for modeling complex, distributed problems [Huhns and Singh, 1998]. Using the AOSE approach, a DC system can be viewed as a network of collaborative, yet autonomous, units modeled as interacting agents that proliferate, regulate, control and organize all activities involved in a distributed, dynamic, and observable environment. Research literature on intelligent agent system architectures has proven that problems that are inherently distributed or require the synergy of a number of distributed elements for their solution can be efficiently implemented as a *multi-agent system* (MAS).

MAS architectures exploit the coupling of the agent effectiveness, acting within a distributed environment, with the advantages of a strict and successfully coordinated framework, as far as communication and agent collaboration is concerned. MAS complexity is solely dependant on the number of agents that participate in it. One could, therefore, argue that the simplest form of MAS consists of a single agent, while more elaborate MAS structures may comprise a substantial number of cooperating agents, or even smaller MAS.

The term "collaboration" does not necessarily imply well-intentioned agent cooperation, but also rivalry, negotiations, and even agent confrontation, always with respect to the desires and intentions of each agent.

The following issues delineate the rationale for MAS over single-agent systems [Agent Working Group, 2000]:

1) A single agent that handles a large amount of tasks lacks performance, reliability, and maintainability. On the other hand, MAS provide modularity, flexibility, modifiability, and extensibility, due to their distributed nature.

2) A single agent cannot obtain (and provide to humans) extensive and specialized knowledge. A MAS, composed of multiple distributed processes can access more knowledge resources, exploiting concurrent execution.

3) Applications requiring distributed computing are better supported by MAS than by a single (often static) agent.

4) Intelligence, as defined by neuroscientists of the human mind, can be approached by a multi-processing system, rather than serial computing. Such a multi-processing computational system can only be implemented as a wide distributed environment where many agents act. Thus, MAS appear as the optimal solution for implementing *Intelligence* in *Intelligent agent-based applications.*

It becomes evident that the choice between a MAS or a single-agent architecture must be guided by the application needs. If we decide, for example, to implement a system notifying us whenever we have an incoming e-mail, then a static single-agent implementation seems sufficient. On the other hand, if we are to implement a large-scale electronic marketplace, where many agents take part in electronic auctions to buy or sell goods, a MAS architecture is warranted.

2.1 Multi-Agent System Characteristics

The core features of multi-agent systems are [Huhns and Stephens, 1999]:

- MAS provide an infrastructure for the specification of communication and interaction protocols
- MAS are typically open and have no central control.
- MAS comprise agents that are scattered around the environment and act either autonomously, or in collaboration

Table 3.1 lists the key properties of an environment with respect to an agent that resides in it [Kaelbling and Rosenschein, 1990].

Table 3.1. Environment characteristics with respect to agents

Property	Definition
Knowable	The extent the environment is known to agents
Predictable	The extent the environment can be predicted by agents
Controllable	The extent the agents can modify the environment
Historical	Whether future states depend on the entire history, or only on the current state
Teleological	Whether parts of the environment are purposeful
Real-time	Whether the environment can change, while agents are deliberating

2.2 Agent Communication

In order for a MAS to solve common problems coherently, the agents must *communicate* amongst themselves, *coordinate* their activities, and *negotiate* once they find themselves in conflict [Green et al., 1997].

In principle, agents communicate in order to achieve the optimal result, either for themselves as individuals, or for the system/society they reside in. Communication enables agents to coordinate their actions and behaviors, therefore resulting in better organized and more coherent systems [Huhns and Stephens, 1999]. The fundamental issues that are related to agent communication are analyzed below [Simon, 1996].

The degree of coordination between agents acting within an environment depends on whether they avoid extraneous activity by reducing contention, avoiding livelock and deadlock, and maintaining applicable safety conditions [Huhns and Stephens, 1999]. Depending on the collaborative and competitive features agents exhibit, coordination is characterized as *cooperation* and *negotiation*, respectively. Figure 3.2 provides a classification of the ways agents can coordinate their behaviors and actions [Weiss, 1999].

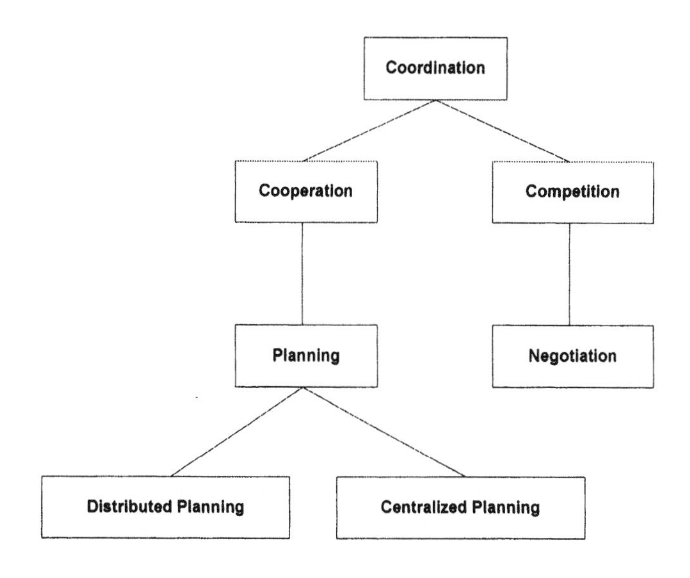

Figure 3.2. Alternative agent coordination manners

Huhns & Singh [Huhns and Singh, 1998] have identified a tri-level perspective on agent communication: a) syntax (how messages are structured), b) semantics (what is the content of the message), and c) pragmatics (how the message is interpreted). Meaning is the combination of semantics and pragmatics [Huhns and Stephens, 1999]. Since communication plays such a pivotal role in MAS efficiency, agents must be well-aware of the different viewpoints of meaning, which are [Singh, 1997]:

1. Descriptive vs. Prescriptive
2. Personal vs. Conventional Meaning
3. Subjective vs. Objective Meaning
4. Speaker's vs. Hearer's vs. Audience Perspective
5. Semantics vs. Pragmatics
6. Contextuality
7. Coverage
8. Identity, and
9. Cardinality

It has now become obvious that agent communication can be defined at any level of abstraction. Nevertheless, it is of great importance to define the lowest of the levels that way, in order to provide communication abilities to even the least capable of the agents in a system and allow them to participate in dialogs. Within the context of a dialog, the agent can be active, passive, or both, functioning as a master, a slave, or peer, respectively.

Agents communicate via messages, which can be of two types: *assertions* and *queries* [Huhns and Stephens, 1999]. In order for an agent to take part in a dialog as a slave, it must be able to accept questions and answer them with assertions. An agent participating in a dialog as a master, must have the ability to make both questions and assertions. Finally, in order for an agent to participate in a dialog as a peer agent, it must possess the abilities of both the master and the slave.

In general, the structure of an agent message is [The FIPA Foundations, 2000]:

1. Message sender
2. Message receiver(s)
3. Language protocol
4. Encoding and decoding functions
5. Actions to be taken

2.3 Agent Communication Languages

Agent communication languages (ACL) are based on *speech acts*, which attempt to model agent communication in a human-like manner [Traum, 1999]. According to speech acts, a sender's request should be stated firmly, in order for the receiver to determine the appropriate message type for its answer. This way the process of designing agent communication is greatly simplified.

In addition to speech acts, ontologies have been been developed in order to improve communication between agents and multi-agent systems. The Webster dictionary [Merriam-Webster, 2000] defines the term *ontology* as a "particular theory about the nature of being or the kinds of existants". In computer science, an ontology is a common lexicon of items, concepts, and their interdependencies. In other words, when two agents are using certain terms in their conversation, it is important that both of them have the same concept regarding these terms. Their vocabulary can be unified through an ontology. This is why the ACLs generally offer the possibility for the sender to indicate which ontology should be used to interpret the content of its message.

Examples of typical ontology representation languages include GOL [Degen et al., 2001], Ontolingua [Gruber, 1992], F-Logic [Kifer et al., 1991] and CyC-L [Lenat, 1995].

The most significant efforts in agent communication, *KQML*, *KIF* and *FIPA-ACL* are described below.

2.3.1 KQML

The Knowledge and Querying Manipulation Language (KQML) [Finin et al., 1997] is a language and a protocol for exchanging information and knowledge. It is part of a larger effort, the ARPA Knowledge Sharing Effort [Patel-Schneider, 1998], which focuses on the development of techniques and methodologies for building large-scale sharable and reusable knowledge bases. KQML is both a message format and a message-handling protocol to support knowledge sharing amongst agents at run-time. It can also be used as a communication language, supporting the interaction of an application with an intelligent system, and enabling knowledge sharing between two or more intelligent systems. KQML is independent of the transport mechanism, the content language and the ontology used, and focuses on the development of an extensible set of performatives [Labrou and Finin, 1998], which define the permissible operations of agents. The performatives comprise a substrate on which higher-level models of inter-agent interaction are developed, i.e., contract nets and negotiation strategies. Finally, KQML provides the infrastructure for knowledge sharing through a special class of agents called communication facilitators, which coordinate the interactions of other agents [Genesereth and Ketchpel, 1994]. KQML has become quite popular, especially in research fields where coordination is of vital importance, like concurrent engineering, intelligent design, and intelligent planning and scheduling. For more information on KQML, one can visit http://www.cs.umbc.edu.

2.3.2 KIF

Knowledge Interchange Format (KIF) is a computer-oriented language for the interchange of knowledge among disparate programs, i.e., a particular logic language, as a standard for describing things within computer systems such as expert systems, databases, intelligent agents, and so on [Labrou et al., 1999]. KIF can be used to support translation from one content language to another, or as a common content language between two agents that use different native representation languages. It provides both a specification for the syntax of a language, as well as a specification for its semantics. Core features include [Genesereth and Fikes, 1992]: i) KIF has declarative semantics (the meaning of expressions in the representation can be understood without appeal to an interpreter for manipulating those expressions); ii) it is logically comprehensive (it provides for the expression of arbitrary sentences in the first-order predicate calculus); iii) it provides for the representation of knowledge; iv) it provides for the representation of non-monotonic reasoning rules, and v) it provides for the definition of objects, functions, and relations. KIF is based on first–order predicate calculus and has a LISP-like prefix syntax, while it is capable of expressing simple facts as well as more complex relationships. For more information on the current specification of KIF, one can visit http://logic.stanford.edu.

2.3.3 FIPA ACL

Although the Knowledge Sharing Effort (KSE) has yielded successful approaches to interoperability problems, it did not offer a disciplined way for developing standards. This void was filled in 1996 by the Foundation for Intelligent Physical Agents (FIPA). FIPA is an international non-profit association of companies and organizations sharing the effort to produce specifications for generic agent technologies. The primary objective of FIPA was to provide specifications that maximize interoperability across agent-based systems, through a set of normative rules that allow a society of agents to exist, operate, and be managed [Bellifemine et al., 2001]. FIPA ACL is one of the developed standards that provides a standard way to package messages, in order to guarantee that the recipient of the message will be able to understand its purpose. It defines a minimal set of message types, called *communicative acts*. Each communicative act is described with both a narrative form and formal semantics based on modal logic. The specification also provides the normative description of a set of high level interaction protocols, including requesting an action, contract net and several kinds of auctions [The FIPA Foundations, 2000]. The FIPA ACL is superficially similar to KQML, with its syntax being identical to that of KQML' s, except for the different names for some reserved primitives. Thus, FIPA ACL, like KQML, does not make any commitment to a particular content language. The current specification of FIPA ACL can be found at http://www.fipa.org.

2.4 Agent Communities

Hillbrand and Stender report that MAS can help us with modeling problems, where individuals exhibit complex behaviors [Hillbrand and Stender, 1994]. MAS can also take into account quantitative, as well as qualitative, system characteristics. By employing a number of communicating agents, simulation of a world of interactions may become feasible. Such MAS structures that evolve in order to encapsulate extremely specialized and complicated behaviors are defined as *agent communities*.

Moreover, a *collaborative community* is an agent group where a) all agents have a common goal and b) each agent is called to serve its part towards the accomplishment of this goal.

According to Ferber [Ferber, 1999], the objectives of an agent community usually are:

- The *validation of a hypothesis* on the emergence of some societal structure. This societal structure should arise from the behaviors and interactions of individuals. Thus, micro-level modeling and parametrization can lead to interesting macro-level phenomena.

- The *development of a theory* on ethological, societal, or psychosocietal aspects of agents, always with respect to their structural and organization characteristics.

- The *integration of theories* that stem from subsidiary research fields like sociology, ethnology, ethology, and psychology, within a unified framework, that provides the appropriate tools.

Apart from agent interactions within a community [Gasser, 1991], [Jennings, 1993], certain rules that govern agent behavior in a community may exist [Rosenschein and Zlotkin, 1994].

PART II

METHODOLOGY

Chapter 4

EXPLOITING DATA MINING ON MAS

1. Introduction

The synergy of two rather diverse technologies, such as DM and AT, has been attempted at many levels. A great number of researchers with varied scientific backgrounds have been involved in the development of systems that use agents to automate data collection while satisfying the imperative need for the interpretation and exploitation of massive data volumes. In fact, each research community defines the basic concepts of AT and DM technologies in a way that increases comprehension and familiarity among its members. This the reason for having summarized the basic DM and AT concepts through the prism of the current work.

As already mentioned in Chapter 1, the main objective of this book is the presentation of a unified methodology for building MAS that have the ability of dynamically incorporating DM-extracted knowledge. Two are the main issues that hinder the coupling of Data Mining with Intelligent Agents:

1. The different logic paradigms that DM and MAS embrace, since DM techniques follow the inductive, while MAS usually adopt the deductive logic paradigm.

2. Both DM and MAS technologies have a wide application range. Thus, categorizing MAS with respect to the type of DM-extracted knowledge is rather complicated.

In this chapter, we discuss the two types of logic and introduce the concepts of agent training and knowledge diffusion, which form the foundation of the AT–DM symbiosis. Then, we define three different levels of knowledge diffusion. The chapter concludes with an overview of MAS development platforms, with an emphasis on Agent Academy.

1.1 Logic and limitations

The increasing demand for sophisticated and autonomous systems, along with the versatility and generic nature of the multi-agent technology paradigm has led to the employment of AT in a variety of disciplines. Requirements of such systems are, among others, the processing of vast amounts of information, the cooperation and coordination of different processes and entities leading to unified solutions, even the development of intelligent recommendations that humans can incorporate into their decisions. Based on this motivation, researchers in the areas of Artificial Intelligence and Software Engineering are oriented towards hybrid approaches that combine different theoretical backgrounds and algorithms. A key prerequisite for the merging of technologies is the existence of a "common denominator" to build upon. In our case, the "common denominator" for AT and DM is the inference procedures they both deploy, which can be expressed by Eq. 4.1:

$$Data \wedge Knowledge| = Information \tag{4.1}$$

As already mentioned, AT and DM assess Eq. 4.1 in two different ways: AT employs deductive logic, whereas DM employs inductive logic.

Deduction is defined as the logical reasoning process in which conclusions must follow their premises. It consists of inferences of the form:

	All Hobbits are Short *Frodo is a Hobbit*
Therefore	*Frodo is Short*

In other words, a knowledge model is applied to data in order to produce information, either in the form of data or knowledge [Fernandes, 2000]. In general, deductive systems are truth preserving, that is, they transform their inputs while conserving their true value. This feature, along with the assumption that initial knowledge is correct, allows the semantic correctness of the knowledge form to be assumed [Talavera and Cortes, 1997].

Usually deduction is used when the process and the goals are well-defined and the human expert, who constructs the knowledge base, has a fine grasp of the problem's underlying principles. Then the system can be modeled appropriately and conclusions can be drawn based on the rules and procedures implemented, satisfying efficiency and soundness.

Nevertheless, deductive rationality has certain limitations, since it breaks down under complexity. On one hand, the logical apparatus ceases to cope beyond a certain complexity, while on the other, in interactive situations of complication, agents may be forced to guess their behaviors. Guessing may lead to a world of subjective beliefs where objective, well-defined assumptions are no longer applicable and therefore the problem becomes ill-defined [Arthur, 1994]. In addition, deductive reasoning systems lack adaptivity, as they are often disembodied from their environment [Wooldridge, 1999].

These drawbacks can be overcome by the use of induction, which is defined as the inference from the specific to the general [Chen, 1999]. Induction produces inferences of the form:

Frodo is a Hobbit, and Frodo is Short
Pippin is a Hobbit, and Pippin is Short
Meriadoc is a Hobbit, and Meriadoc is Short
...

Therefore *All Hobbits are Short*

Induction follows a different approach towards the realization of Eq. 4.1. Data and information are known, (data tuples and their classification, respectively) and the objective is to extract the knowledge model that "transforms" the former into the latter. The main primitives of induction are the discovery of previously unknown rules, the identification of correlations, and the validation of hypotheses in available datasets. On the negative side, the discovered knowledge may not always be valid, since inductive learning systems transform their inputs by means of progressive generalizations [Kodratoff, 1988]. Therefore, induction alone is unable to satisfy system soundness.

One may safely argue that neither of the above approaches is a panacea in real-life applications. Even though both inductive and deductive reasoning have been used separately in a variety of fields, we believe that the real coupling of these two practices (through the coupling of their representative technologies) may lay the foundation for enhanced, more efficient, and market-oriented MAS. In our approach, we exploit the power of deductive reasoning for facilitating the well-known and established procedures and functionalities of the MAS. We also use inductive reasoning for the discovery and exploitation of previously unknown knowledge, a process equivalent to the adaptation of the MAS to real-world, real-time requirements.

1.2 Agent Training and Knowledge Diffusion

In order to build efficient MAS, the confluence of agent modeling and knowledge extraction is a prerequisite. In fact, every agent with reasoning capabilities in a MAS must have the internal structure illustrated in Figure 4.1.

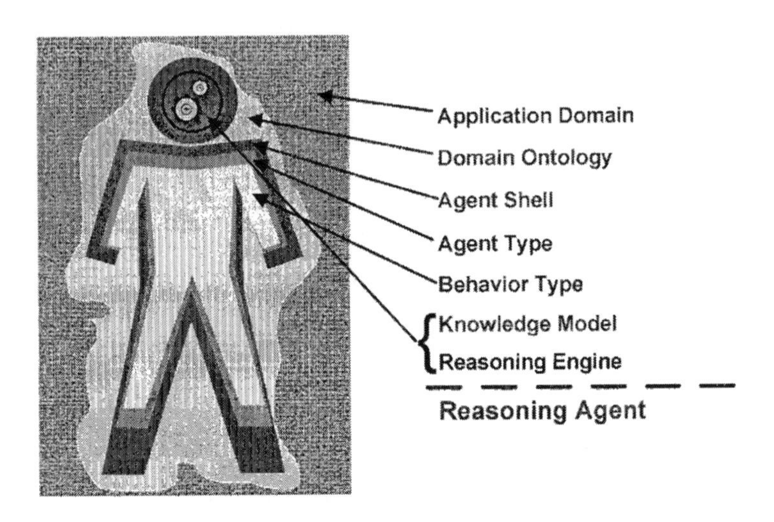

Figure 4.1. The structure of the reasoning agent

The various layers listed in the Figure correspond to the different building blocks produced by associated processes. Construction of an agent usually begins from the outside in. The first four (outer) layers of domain ontology, agent shell, agent type, and behavior type are typically determined using widely available MAS development tools. Some of these tools and platforms are presented in Section 2 of this chapter.

The last and most interesting and complicated task involves the creation of the knowledge model. We argue that this model can be generated through data mining and inserted to the, otherwise, "dummy" agent.

The process of dynamically incorporating DM-extracted ***knowledge models*** (KMs) to agents and MAS is defined as ***training***, while the process of revising, in order to improve, the knowledge model(s) of agent(s), by re-applying DM techniques is defined as ***retraining***. We, finally, define the concept of ***knowledge diffusion***, as the outcome of the incorporation of DM-extracted knowledge to agents.

Different types of extracted knowledge, can lead to different knowledge diffusion levels, providing a criterion for MAS categorization.

1.3 Three Levels of Knowledge Diffusion for MAS

Following the above perspective, data mining techniques can be applied in three alternative manners, leading to three different types of knowledge, which, in turn, correspond to three distinct ways of knowledge diffusion to the MAS architecture:

1 *Data mining on the application level of a MAS*
Data mining is performed on available application data, in order to discover useful rules and/or associations and/or patterns. The extracted knowledge is related to the scope of the end-user application and not its internal architecture. This approach is suitable for applications that are not yet agent-based, but have a wealth of historical data. In this case, the technology coupling issue is viewed macroscopically, where the knowledge models extracted are intended to improve the efficiency of the MAS.

2 *Data mining on the behavior level of a MAS*
In this case, data mining is performed on log files containing agent behavior data (i.e., actions taken, messages exchanged, decisions made). The objective is to better predict future agent behaviors by eliminating unnecessary or redundant agent activity. The extracted knowledge may result in more efficient agent actions and thus, reduce system workload. Naturally, this approach is applicable only when the end-user application is already agent-based. In this case, the coupling issue is viewed microscopically, where the knowledge models extracted are intended to improve the performance of an agent engaged in a MAS (i.e., to improve the internal MAS action flow).

3 *Data mining on evolutionary agent communities*
At this level, we perform evolutionary data mining techniques on agent communities, in order to study agent societal issues. This approach has to do with the satisfaction of the goals of an agent community, which evolves and learns through interaction. In this case, coupling is considered at a higher level of abstraction, where the main focus is on the formulation of the problem that the agent community is confronted with and the way the extracted knowledge is diffused to the agent community.

2. MAS Development Tools

The growth in interest and the use of agent technology has motivated the development of different frameworks and environments that aim at facilitating rapid design and implementation of MAS. Most of them are Java-based applications that provide mechanisms to manage and monitor all issues concerning agent design, communication, coordination, and control. Such frameworks include [Shen et al., 2001; Eiter and Mascardi, 2002; Mangina, 2002]:

- **Aglets**
  Aglets is a Java mobile agent platform and library that eases the development of agent based applications. An aglet is a Java agent able to autonomously and spontaneously move from one host to another, taking along its code, as well as its data. Aglet communication is established through a whiteboard, allowing both synchronous and asynchronous agent collaboration. The Aglets technology is now hosted at sourceforge.net as an open source project (http://aglets.sourceforge.net/).

- **JAFMAS/JIVE**
  JAFMAS (Java-based Agent Framework for Multi-Agent Systems) provides a framework to guide the coherent development of multiagent systems along with a set of classes for agent deployment in Java. It focuses on agent development from a speech-act perspective. The framework enables both directed as well as subject-based broadcast communication. JAFMAS is also developed in Java and is available at http://www.ececs.uc.edu/~abaker/JAFMAS.

- **JATLiteBean**
  The JATLiteBean is a JavaBean component encapsulating and extending the functionality of the JATLite (Java Agent Template, Lite) agent toolkit, a package of programs that allow users to quickly create new software agents that communicate robustly over the Internet. JATLite does not, by itself, construct "intelligent agents", rather it facilitates the construction of agents, particularly those that communicate by sending and receiving messages using KQML (http://www.cs.umbc.edu/kqml/ for the current KQML standard). The communications are built on open Internet standards, i.e., FTP, TCP/IP, and SMTP. However, developers may easily build agent systems using other agent languages, such as the FIPA ACL using the JATLite layers. JATLiteBean is available at http://kmi.open.ac.uk/people/emanuela/JATLiteBean/.

- **SIM_AGENT**
  The Sim_Agent toolkit provides a range of resources for research and teaching related to the development of interacting agents in environments of various degrees and kinds of complexity. Work is mainly

focused on exploring architectural design requirements for intelligent human-like agents, so Sim_Agent is not committed to a particular type of agent architecture. It is a free, open source toolkit for designing agents with hybrid architectures. Sim_Agent supports rapid prototyping, incremental compilation, and changes to code without having to restart a process. It also provides support for a wide variety of AI techniques and programming styles, and the tools for graphical tracing via displays of a running simulation and graphical interaction via a mouse. Sim_Agent is available at http://www.cs.bham.ac.uk/~axs/cog_affect/sim_agent.html.

- **ZEUS**
  Zeus is a "collaborative" agent building environment and component library written in Java. Each ZEUS agent consists of a definition layer, an organizational layer and a co-ordination layer. The definition layer represents the agent's reasoning and learning abilities, its goals, resources, skills, beliefs, preferences, etc. The organization layer describes the agent's relationships with other agents. The co-ordination layer describes the co-ordination and negotiation techniques the agent possesses. Communication protocols are built on top of the co-ordination layer and implement inter-agent communication. Beneath the definition layer is the API. ZEUS is available at http://www.labs.bt.com/projects/agents/zeus/.

- **FIPA-OS**
  FIPA-OS is a component-based toolkit enabling rapid development of FIPA compliant agents. It comprises the following components: the Agent Shell, the Task Manager, the Conversation Manager, the Message Transport Service, the JESS agent Shell, the Database Factory, and the Parser Factory. FIPA-OS supports the majority of the FIPA Experimental specifications and is being continuously improved as a managed Open Source community project, making it an ideal choice for any FIPA compliant agent development activity. FIPA-OS was first released in August 1999 as the first publicly available, royalty-free implementation of FIPA technology. Available at http://fipa-os.sourceforge.net/index.htm

- **JADE**
  JADE (Java Agent Development Framework) is a software framework fully implemented in Java language. It simplifies the implementation of multi-agent systems through a middleware that complies with the FIPA specifications and through a set of graphical tools that supports the debugging and deployment phases. The agent platform can

be distributed across machines (which do not even need to share the same OS) and the configuration can be controlled via a remote GUI. The configuration can be changed at run-time by moving agents from one machine to another, as and when required. JADE is entirely implemented in Java and the minimal system requirement is Version 1.4 of the language (the run time environment or the JDK). The efficient synergy between the JADE platform and the LEAP libraries has resulted in a FIPA-compliant agent platform with reduced footprint and compatibility with mobile Java environments down to J2ME-CLDC MIDP 1.0. Both JADE and LEAP libraries are available at http://jade.tilab.com/.

Although efficient, none of the existing development platforms can embed knowledge *automatically* and *dynamically* into the agents of a MAS. This is why we have developed *Agent Academy* [Mitkas et al., 2002; Mitkas et al., 2003], an integrated framework for constructing multi-agent applications and for embedding rule-based reasoning into agents, both at the design phase and at runtime. A description of the framework follows.

3. Agent Academy

Agent Academy (AA) has been implemented upon the JADE infrastructure and is available at http://www.source-forge.net/projects/Agent academy. AA is itself a multi-agent system, whose architecture is based on the GAIA methodology [Wooldridge, 1997]. It constitutes an integrated GUI-based environment that supports the design of both single agents and multi-agent systems, using common drag-and-drop operations. This capability of AA enables agent application developers to build a whole community of agents with predefined behavior types and attributes within a short time interval. Using AA, an agent developer can easily go into the details of the designated behaviors of agents and precisely regulate their communication properties. These include the type and number of the ACL messages exchanged between agents, the performatives and structure of messages –always conforming to FIPA specifications [The FIPA Foundations, 2000]–, as well as the semantics, which can be defined by constructing ontologies with Protégé-2000 [Grosso et al., 1999].

Developing an agent application using AA involves the following activities from the developer's side:

a. the creation of new agents with limited initial reasoning capabilities;

b. the addition of these agents into a new MAS;

c. the determination of existing, or the creation of new behavior types for each agent;

d. the importation of ontology-files from Protégé–2000;

e. the determination of message recipients for each agent.

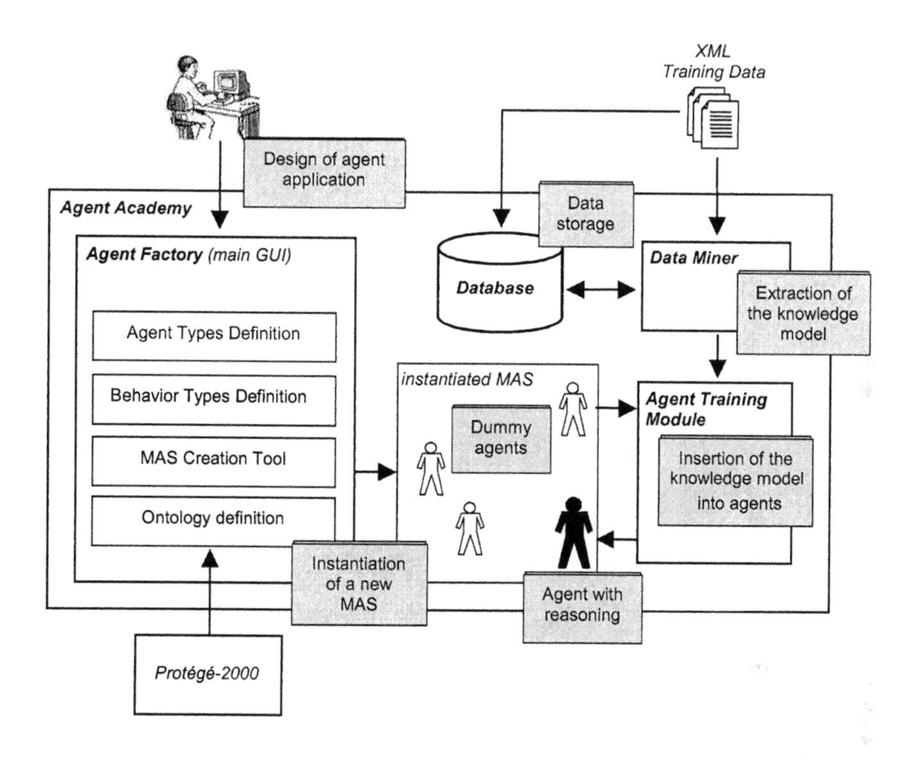

Figure 4.2. Diagram of the Agent Academy development framework

Figure 4.2 illustrates the Agent Academy functional diagram, which includes the main components and the interactions between them. In the remainder of this Chapter we review the Agent Academy architecture and explain how the entire development process, except agent training, is realized through our framework. Agent training issues are thoroughly analyzed in the next Chapter, while retraining issues are discussed in Chapter 9.

3.1 AA Architecture

An application developer launches the AA platform in order to design a multi-agent application. The main GUI of the development environment is provided by the *Agent Factory*, a specifically designed agent, whose role is to collect all required information from the agent application developer. Such information includes the definition of the types of agents involved in the MAS, the types of behaviors of these agents, as well as the ontology they share with each other. For this purpose, Agent Academy is equipped with a Protégé–2000 front-end. The initially created agents possess no referencing capabilities ("dummy"agents). The developer may request from the system to train one or more agents of the new MAS. These agents interoperate with the *Data Miner*, which is responsible for producing knowledge models by the use of DM techniques and for inserting them (with the help of the *Agent Training Module*) into the agents. A detailed description of this process is provided in the next Chapter.

Finally, it should be noted that AA hosts a database for storing all information about the configuration of the new created agents, their knowledge models, as well as data entered into the system for DM purposes.

3.2 Developing Multi-Agent Applications

The main GUI of AA (Agent Factory) offers a set of graphical tools, which enable the developer to carry out all required tasks for the design and creation of a MAS, without any effort for writing even a single line of source code. In particular, the Agent Factory comprises the Ontology Design Tool, the Behavior Type Design Tool, the Agent Type Definition Tool, and the MAS Creation Tool.

3.3 Creating Agent Ontologies

A required process in the creation of a MAS, is the design of one or more ontologies, in order for the agents to interoperate. The Agent Factory provides an *Ontology Design Tool*, which helps developers adopt ontologies defined with the Protégé–2000. The RDF files that are created with Protégé are saved in the AA database for further use.

3.4 Creating Behavior Types

The *Behavior Type Design Tool* assists the developer in designing and developing generic behavior templates. Agent behaviors are modeled as workflows of basic building blocks, such as receiving/sending a message, executing an in-house application, and, if necessary, deriving decisions using inference engines. The data and control dependencies between these blocks are also handled. The behaviors can be modeled as *cyclic* or *one-shot* behaviors of the JADE platform. These behavior types are generic templates that can be configured to behave in different ways; the structure of the flow is the only process defined, while the configurable parameters of the application inside the behavior, as well as the contents of the messages can be specified using the MAS Creation Tool. It should be denoted that the behaviors are customized to the application domain.

The building blocks of the workflows, which are represented by nodes, can be of four types:

- *Receive nodes*, which enable the agent to filter incoming FIPA-SL0 messages.
- *Send nodes*, which enable the agent to compose and send FIPA-SL0 messages.
- *Activity nodes*, which enable the developer to add predefined functions to the workflow of the behavior, in order to permit the construction of multi-agent systems for existing distributed systems.
- *Jess nodes*, which enable the agent to execute a particular reasoning engine, in order to deliberate about the way it will behave.

Figure 4.3 illustrates the design of the behavior for an agent that receives a message and, according to the content of the message, either executes a pre-specified function, or sends a message to another agent.

3.5 Creating Agent Types

Having defined certain behavior types, the *Agent Type Definition Tool* is called to create new agent types, which will be later used in the *MAS Creation Tool.* An agent type is in fact an agent plus a set of behaviors assigned to it. New agent types can be constructed from scratch or by modifying existing ones. Agent types can be seen as templates for creating agent instances during the design of a MAS.

During the MAS instantiation phase, which is realized by the use of the MAS Creation Tool, several instances of already designed agent types will be generated, with different values for their parameters. Each agent instance of the same agent type can deliver data from different data sources, communicate with different types of agents, and even execute different reasoning engines.

3.6 Deploying a Multi Agent System

The design of the behavior and agent types is followed by the deployment of the MAS. The *MAS Creation Tool* enables the instantiation of all defined agents running in the system from the designed agent templates. The receivers and senders of the ACL messages are set in the behaviors of each agent. After all the parameters are defined, the agent

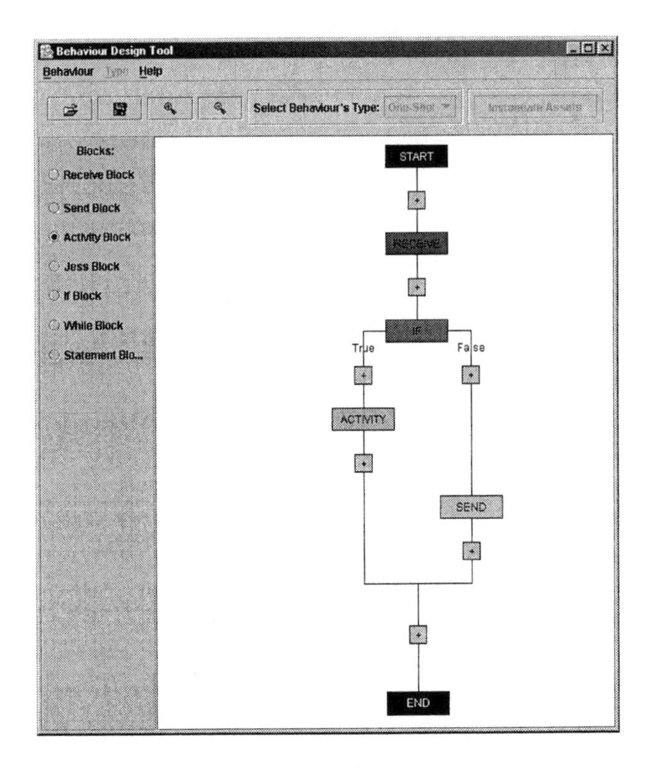

Figure 4.3. Creating the behavior of an agent through the Behavior Design Tool

instances can be initialized. Agent Factory creates *default AA Agents*, which have the ability to communicate with the components of AA, and augments them with the necessary ontologies and behaviors.

Now, agents are ready for some serious training!

Chapter 5

COUPLING DATA MINING WITH INTELLIGENT AGENTS

In this chapter, we present a unified methodology for MAS development, which relies on the ability of DM to generate knowledge models for agents. The basic methodology encompasses a number of stages and is suitable for all three cases of knowledge defined in Chapter 4. Most of the steps are common to all cases, while some require minor modifications to accommodate each case (see Figure 5.1). The second half of this chapter provides a description of the Data Miner, the tool that we have developed in an effort to automate a large portion of the agent training and retraining processes.

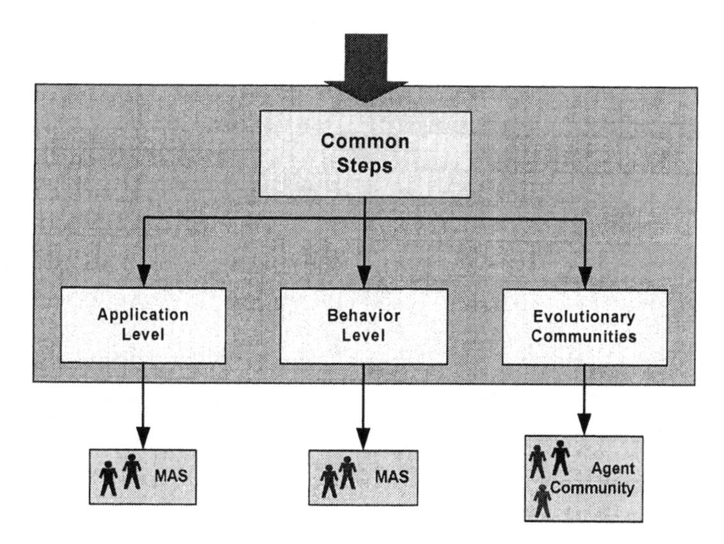

Figure 5.1. The unified MAS methodology

1. The Unified Methodology

1.1 Formal Model

Let O be the ontology of the MAS. Let $A = \{A_1, A_2, \ldots, A_n\}$ be the set of attributes described in O and defined on $\mathcal{D}$, the application data domain. Let $D \subseteq \mathcal{D}$ be a set of data, where each dataset tuple is a vector $t = \{t_1, t_2, \ldots, t_n\}$, and t_i, $i = 1 \ldots n$, is a value for the corresponding attribute A_i. Missing values are allowed within t.

Let us now consider a MAS with k different agent types Q_i, $i = 1, \ldots, k$. For each type, q_i agent instances $Q_i(j)$ may exist with $1 \leq q_i \leq m$. Thus, the total number of agents in the MAS is equal to: $\sum_{i=1}^{k} q_i$. For these agents, three are the training cases, with respect to the knowledge types extracted:

1.1.1 Case 1: Training at the MAS application level

Agents of the same type employ a common knowledge model obtained by performing DM on an appropriate subset of D. This subset contains only the attributes that are relevant to the specific agent type. We, therefore, define $D_{IAQ_i} \subseteq D_{AT}$, ($A \longrightarrow$ *Application*), as the initial training dataset for agent type Q_i, where D_{AT} is the initial application dataset. In most cases $D_{AT} = D$.

1.1.2 Case 2: Training at the MAS behavior level

In order to train agent type Q_i of the MAS at the behavior level, we use a dataset that contains historical data on the actions of the agents of this type. This case requires the existence of a MAS operating for a period long enough to generate sufficient behavior data. We define D_{IBQ_i} as the initial training dataset for agent type Q_i, where $D_{IBQ_i} \subseteq D_{BT}$, ($B \longrightarrow$ *Behavior*) and D_{BT} is the initial behavior dataset.

It should be noted that the availability of an initial dataset on agent behaviors is not very likely and, consequently, initial agent training may not be common. Nevertheless, agents may be equipped with simple, deductive rules in the beginning, and data mining can be applied only after a substantial number of agent action records has been collected. An agent action monitoring and recording mechanism is, thus, considered of great importance.

1.1.3 Case 3: Training evolutionary agent communities

Finally, in order to train agent type Q_i, residing in an agent community, instead of exploiting an initial dataset, we formulate the agent logic model in a way that allows the application of evolutionary DM techniques. Once more we define D_{IEQ_i} ($E \longrightarrow$ *Evolutionary*), as the initial agent training model for agent type Q_i, where $D_{IEQ_i} \subseteq D_{ET}$ and D_{ET} is the set of initial training models of the agents.

In order to provide a unified framework for the training and retraining concepts, we also define the datasets D_{IQ_i} and D_T, as the generic forms of the abovementioned training cases ($D_{IAQ_i}/D_{IBQ_i}/D_{IEQ_i}$ and $D_{AT}/D_{BT}/D_{ET}$, respectively).

1.2 Common Primitives for MAS Development

Figure 5.2 illustrates the unified methodology for developing MAS that can exploit knowledge extracted by the use of DM techniques. On the one hand, standard AOSE processes are followed, in order to specify the application ontology, the agent behaviors and agent types, the communication protocol between the agents, and their interactions. On the other hand, DM techniques are applied for the extraction of appropriate knowledge models.

The methodology pays special attention to two issues: a) the ability to **dynamically** embed the extracted KMs into the agents and b) the ability to repeat the above process as many times as deemed necessary.

The ten steps of the methodology are listed below. They are also illustrated in the form of a flow diagram in Figure 5.3.

①. Develop the application ontology

②. Design and develop agent behaviors

③. Develop agent types realizing the created behaviors

④. Apply data mining techniques on the provided dataset

⑤. Extract knowledge models for each agent type

⑥. Create the agent instances for the application

⑦. Dynamically incorporate the knowledge models to the corresponding agents

⑧. Instantiate multi-agent application

⑨. Monitor agent actions

⑩. Periodically retrain the agents of the system

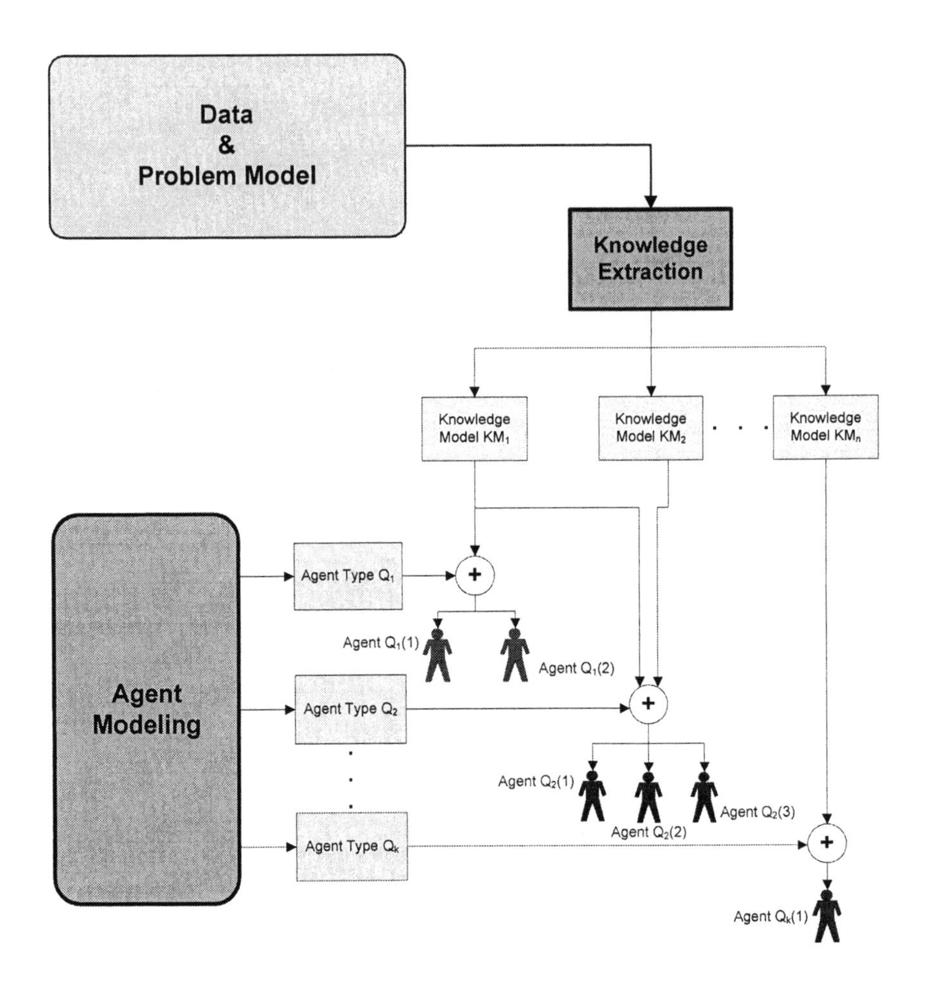

Figure 5.2. The MAS development mechanism

Following this methodology, we can automate (or semi-automate) several of the processes involved in the development and instantiation of MAS. Taking a Software Engineering perspective, we can increase the adaptability, reusability, and flexibility of multi-agent applications, simply by re-applying DM on different datasets and incorporating the extracted KMs into the corresponding agents. Thus, MAS can be considered as efficient add-on components for the improvement of existing software architectures.

Slight variations and/or additions may be needed to the presented methodology in order to adapt it to each one of the three cases. These differentiations may be either related to the agent behavior creation

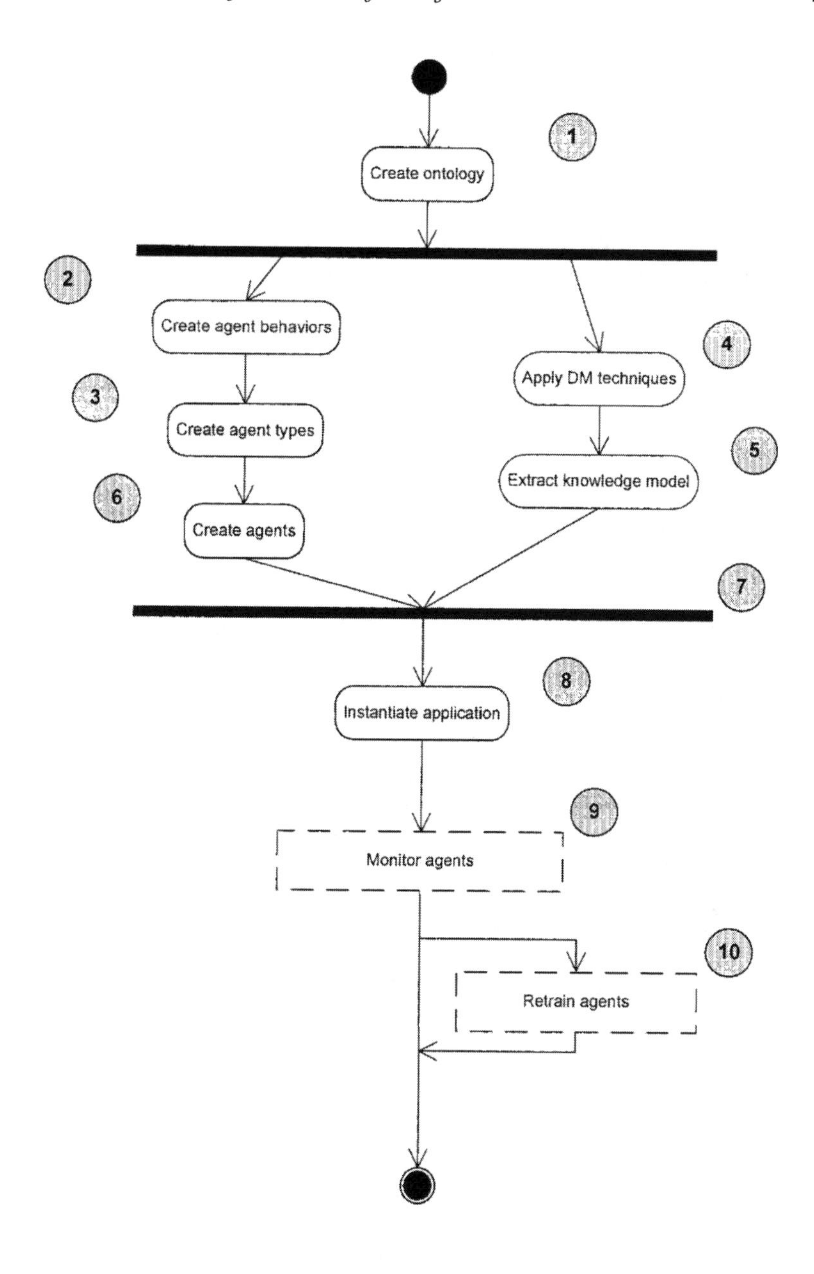

Figure 5.3. The common MAS development steps

phase (Cases 2 & 3 - e.g., model agent actions), the training phase (Cases 1, 2, & 3 - e.g., training set selection, knowledge model representation, retraining intervals), or the retraining phase (Cases 1,2, & 3 - e.g., retraining initiation). We now discuss each case separately.

1.3 Application Level: The Training Framework

Let us consider a company that would like to embrace agent technology and transform its processes from legacy software to a MAS. During the modeling phase, agent roles and interactions are specified. Then, agent behaviors and logic models have to be defined, a process extremely complicated, since the specification of an appropriate (sound) set of rules is not always straightforward.

In most cases, domain understanding is infused to the system in the form of static business rules, which aim to satisfy refractory, nevertheless suboptimal, MAS performance. In the case the rules are set by a domain expert, there is a high probability of ignoring interesting correlations between events that may occur during system operation. Moreover, the rigidity of the rules introduced, cannot "capture" the dynamism of the problem at hand (the problem the MAS is assigned to solve). The extraction of useful knowledge models from historical application data is, therefore, considered of great importance.

In general, MAS that aim to improve the internal processes of a company, have to follow certain primitives, in order to enable control and facilitate knowledge diffusion. Such MAS should be organized into four discrete, yet closely related layers, where the flow of data and meta-data is unrestricted (Figure 5.4).

Table 5.1 illustrates the basic functionalities of each layer.

Table 5.1. The basic functionalities of each layer

Layer	Functionalities
Middleware	- Input data - Preprocess data - Select data
Knowledge Discovery	- Apply DM technique - Tune DM algorithm - Extract knowledge model
Knowledge Organization	- Evaluate & validate knowledge models - Incorporate static business rules - Organize and integrate knowledge
Knowledge Diffusion	- Visualize results - Diffuse information to interested parties

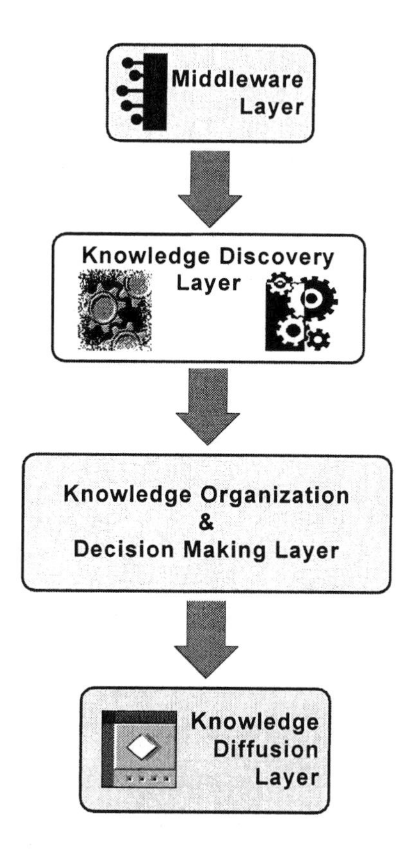

Figure 5.4. Application level: the training framework

1.4 Behavior Level: The Training Framework

The development of a methodology for predicting the actions of agents operating in a MAS, based on their former behavior, is an intriguing issue. Going briefly through related bibliography, one can easily identify a lack of publications on agent action identification and prediction, since it is inherently complicated. However, this problem is very similar to web personalization, where a significant number of research efforts have been published. By drawing an analogy to web personalization, agent behavior prediction may be feasible.

Rather popular approaches for web personalization include collaborative filtering [Konstan et al., 1997; Herlocker et al., 1999] and Web usage mining [Nasraoui et al., 1999; Spiliopoulou et al., 1999; Perkowitz and Etzioni, 1998]. Advancing to more elaborate infrastructures, web usage mining systems have been built [Cooley et al., 1999; Zaïane et al., 1998]

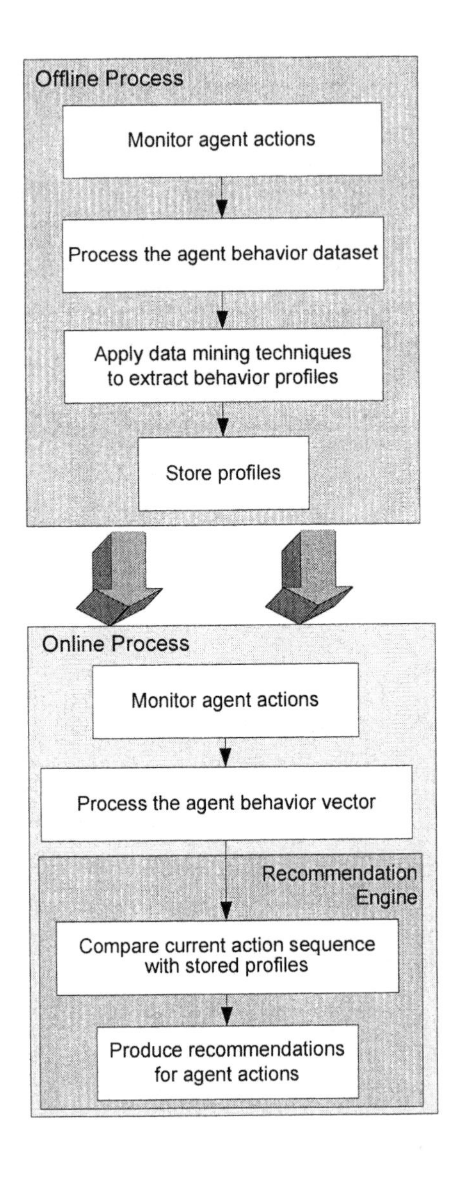

Figure 5.5. The basic functionality of an agent prediction system

to discover interesting patterns in the navigational behavior of users [Mobasher et al., 2000a]. Nevertheless, none of the above approaches had proven sufficient for providing successful personalization of users, and, correspondingly, of agents, until aggregation usage profiles were introduced. Several research groups [Shahabi et al., 1997; Schechter et al., 1998; Banerjee and Ghosh, 2001] adopted this approach to identify association rules [Agrawal and Srikant, 1994; Agarwal et al., 2001], sequential patterns, pageview clusters and transaction clusters between users [Shahabi et al., 1997]. One of the most popular models for personalization and prediction, though, is the one proposed by Mobasher [Mobasher, 1999] and Mobasher, Cooley, & Srivastava [Mobasher et al., 1999; Mobasher et al., 2000b]. Their model has been adopted for developing the agent behavior prediction methodology presented in this book.

The first task in such prediction systems is the collection of all necessary historical data. One may argue that, in order to predict the behavior of an agent, only information on its previous actions is needed. This is not necessarily the case, though, when creating behavior profiles. These profiles, which are extracted by the use of DM techniques, are the projection of "agent experience" on the data.

Depending on the dataset employed for predicting agent actions and for creating profiles, alternative architectures may occur for the prediction system. In a MAS, for example, each agent $Q_i(j)$ could maintain a record for its past actions (dataset $D_{Q_i(j)}$), and use it as a pool of knowledge in order to predict its future actions. In such a system, the predictions would be solely based on each agent's own"experience"; an agent with no (limited) historical record, would, thus, produce no (poor) predictions. An alternative architecture could allow the historical records of all agents (of similar type) to be monitored by an appointed agent and form the dataset D_{Q_i}. Profiles would then be created on the whole of the agent action history, discarding nevertheless personalization. We believe that a system for predicting agent actions should take the middle ground and allow both personalization and collaboration between agents. In this way, if a specific agent cannot create good predictions, it can exploit other agents' experience.

In general, the main goal of a prediction system is the improvement of performance of a wider framework, which specifies the exact development principles and agent interactions of the prediction system.

Figure 5.5 illustrates the functional architecture of such a system. It comprises two modules: a) an offline module, where action preprocessing, DM application, and profile creation takes place and, b) an online module, where the agent actions are recorded in real-time and the recommendation interface is applied.

Taking all these factors into account, a methodology for building a framework capable of predicting agent actions should comprise the following steps:

①. Model agent actions

②. Develop a mechanism for monitoring agent actions

③. Preprocess data, in order to incorporate domain understanding

④. Develop an appropriate DM algorithm for extracting agent action profiles

⑤. Develop a mechanism for storing and retrieving profiles

⑥. Develop an agent action recommendation interface

1.5 Evolutionary Communities Level: The Training Framework

Agent communities are usually deployed to simulate complicated, heterogeneous and non-linear problems, which cannot be efficiently handled by dealt with by the use of multi-agent architectures. In contrast to the former two levels, the internal structure of agent communities is inherently complicated, due to continually varying interactions. In addition, communities have collective goals and their progress towards achieving them cannot be evaluated at the individual agent level. Thus, the performance and efficiency of the agent community must be measured indirectly through the use of suitably selected global indicators.

Developing a generic framework for building agent communities appears to be a daunting task, complicated by the existence of multiple alternatives. The process can be made less difficult if viewed through the perspective of the presented approach, which focuses on the way evolutionary techniques can be exploited to augment agent community intelligence. Certain prerequisites have to be satisfied. First of all, the problem has to be modeled appropriately, in order to be simulated. All the parameters and their impact have to be specified, both for the environment the agents reside, as well as for the agents themselves. In addition, the knowledge model the agents adopt has to be created. It should be noted, that evolutionary DM techniques operate on an existing knowledge model which they try to improve. Proper model realization leads to successful agent training. Another issue of great importance is agent communication, since agent interaction –most of the times implicit– plays a pivotal role in community evolution. As far as the extracted knowledge is concerned, all agents are equipped with the mechanism shown in Figure 5.6, that evaluates agent decisions and adjusts the knowledge model accordingly. Finally, a set of suitable indicators have to be defined, in order to monitor evolution and draw conclusions. The set may combine quantitative and qualitative indicators.

Summarizing, the process of designing and developing an agent community involves the following steps:

①. Model the problem

②. Represent the agent knowledge model

③. Specify the agent communication framework

④. Develop a knowledge evaluation mechanism

⑤. Define the set of community indicators

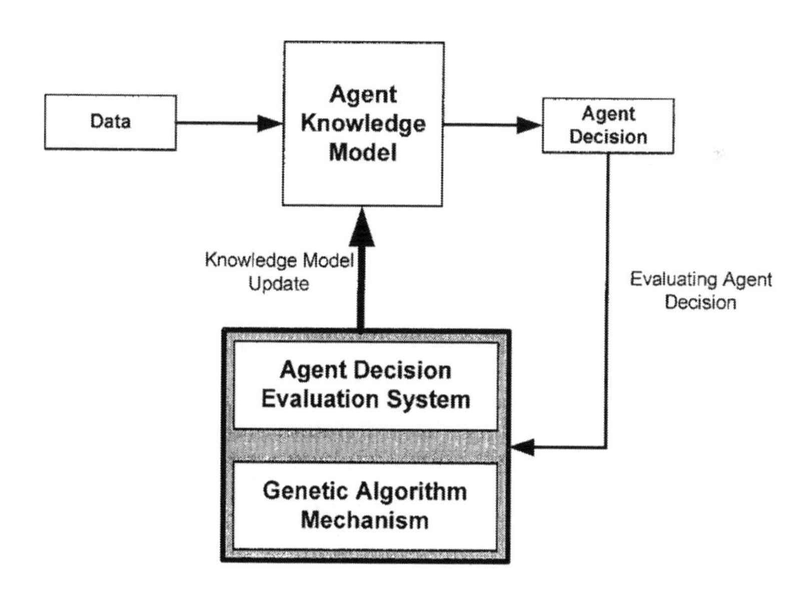

Figure 5.6. The knowledge evaluation mechanism

2. Data Miner: A Tool for Training and Retraining Agents

In order to enable the incorporation of knowledge into agents, we have implemented a semantically–aware DM tool that is agent-oriented. The Data Miner, as its name implies, is a suite that provides users with the capability of applying a number of DM algorithms on application-specific and agent-behavior-specific data. The extracted decision models can then be inserted into JADE agents, augmenting that way their intelligence. The Data Miner is one of the core components of the Agent Academy platform discussed in Chapter 4. It can also function as a stand-alone tool for classification, association rule extraction, and clustering. A stable version of the Data Miner, along with user documentation can be found at: http://sourceforge.net/projects/aadataminer.

2.1 Prerequisites for Using the Data Miner

During the design phase of a MAS, the application developer is called to identify how DM techniques can be exploited, in order to enhance the system under development. The proper technique has to be selected, with respect to the problem at hand and the data available. Lack of data, data inconsistency or unsuitability of data may lead to poor results with the Data Miner.

Next, the developer has to decide which of the MAS agents will be able to incorporate and diffuse DM-extracted knowledge. Not all agent types have to bear the inductive reference mechanism. Data Miner assumes that this mechanism provides agents with the ability to infer and act when exposed to some unknown situation, based on their knowledge and perception of the environment. The term *agent perception* refers to all values of internal variables, content of messages or states that an agent can sense.

2.2 Data Miner Overview

The mechanism adopted for embedding rule-based reasoning capabilities into agents is illustrated in Figure 5.7.

The Data Miner comprises three modules:

1. The Preprocessing Unit, which receives and conditions the input data.

2. The Miner, which is the core component, where data mining takes place.

3. The Evaluator, which is the front-end of the Data Miner and provides algorithm performance metrics and visualization capabilities.

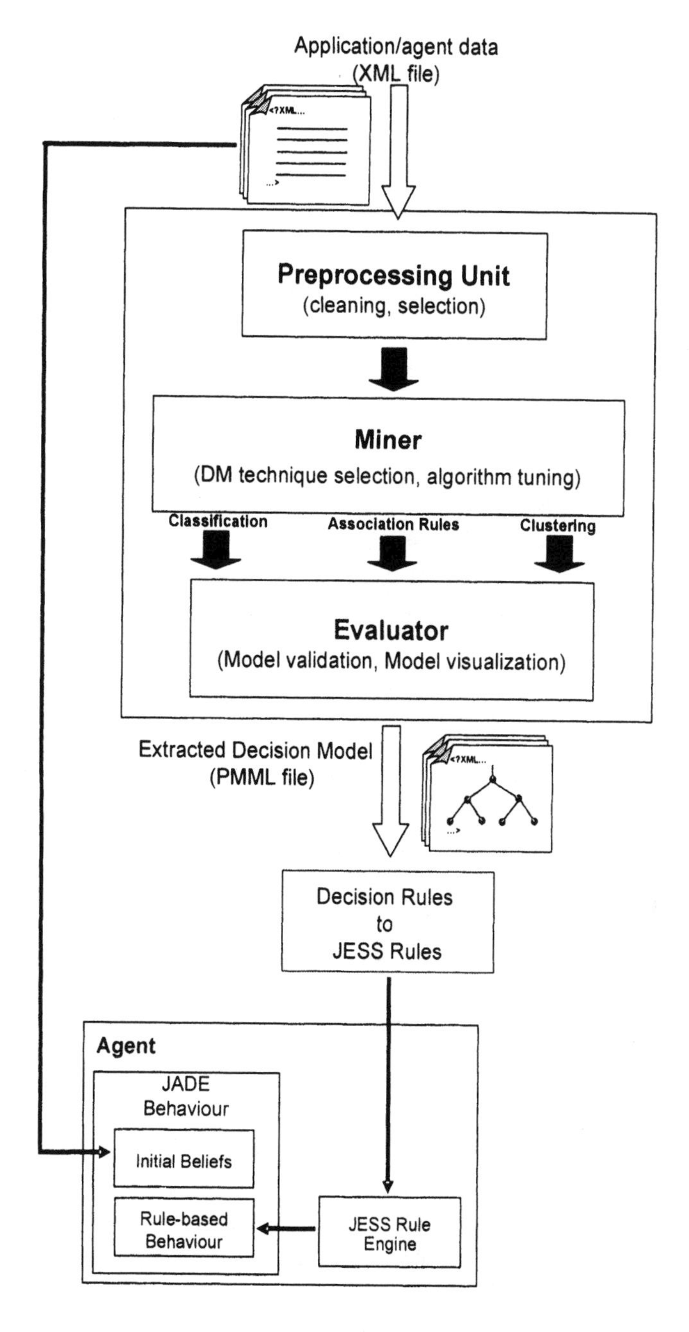

Figure 5.7. The training/retraining mechanism

Data, either application-specific or agent-behavior-specific, are input in XML format. Each data file contains information on a) the name of the agent the file is, or will be assigned to and b) the knowledge model of the agent type the file will be used for.

The XML file is then inserted into the *Preprocessing Unit* of the Data Miner, where all the necessary feature extraction, data cleaning, selection, and reduction tasks take place. Next, data are forwarded to the *Miner*, where the user has to decide on which DM technique and algorithm to employ. Making the proper choice is of vital importance for the usefulness and quality of the extracted knowledge[1].

Table 5.2. Techniques and algorithms provided by the Data Miner

DM technique		
Classification	**Association Rules**	**Clustering**
ID3	Apriori	K-Means
C4.5	DHP	PAM
CLS	DIC	EM
FLR	κ-Profile	-

Table 5.2 illustrates the DM techniques and the associated algorithms that Data Miner offers. Two new algorithms are introduced: FLR, a classification algorithm that produces fuzzy lattice rules [Athanasiadis and Mitkas, 2004], and κ-Profile, an algorithm that perform intelligent segregation of roaming attitudes for predicting agent behaviors (see Chapter 7).

After DM is performed, the results are sent to the *Evaluator*, which is responsible for the validation and visualization of the extracted model. If the user accepts the model, it is transformed into a document written in PMML. PMML, or Predictive Model Markup Language [Data Mining Group, 2001], is an XML-based language that enables sharing of statistical and DM models among different applications. This PMML document expresses the referencing mechanism of the agent we intend to train and provides compatibility and portability. The resulting knowledge model is then translated to a set of facts executed by a rule engine. The implementation of the rule engine is done through the Java Expert System Shell (JESS) [Friedman-Hill, 1998], which is a robust mechanism for executing rule-based agent reasoning. The execution of the rule en-

[1]Section 2.3 below provides some hints for selecting the appropriate technique.

gine transforms the Data Miner extracted knowledge into a living part of the agent's behavior.

2.3 Selection of the Appropriate DM Technique

The process of selecting the best data mining technique for the problem at hand is complex, since such a decision is highly dependent on a number of factors, including data semantics, data quantity and quality, and the application domain itself. Nevertheless, the most important factor is the reason for performing data mining. What is the user looking for in the data? If this question is answered, things get somewhat less complicated.

In this section, we provide some hints on when the user should apply association rule extraction, classification, or clustering, since these are the techniques supported by the Data Miner. It should be denoted, though, that these are only suggestions. No "panacea" exists.

Association rule extraction is recommended when the user wants to:

a) Discover strong relationships between events

b) Identify traversal patterns, and

c) Understand how changing one (or more) variable(s) can affect another

The main application domains of ARE include market-basket and loss-leader analysis, cross-marketing, and catalog design.

Classification, on the other hand, is recommended in order to:

1) Reliably apply the segmentation scheme to a set of data that reflects a group of potential customers

2) Identify possible interactive relationships between variables in a way that would lead to an improved understanding of how changes in one variable can affect another

3) Generate a visual representation of the relationship between variables in the form of, i.e., a tree, which is a relatively easy way to understand the nature of the data residing in a database

4) Simplify the mix of attributes and categories in order to retain the essential ones for making predictions

5) Identify important variables in a data set that can eventually be used as a target.

Classification is applied to any domain where prediction or categorization is desired. We have to remember, though, that the existence of predefined classes is a prerequisite.

Finally, **clustering** is used when:

a. Domain knowledge is minimal and the user desires to learn something about the input parameters

b. A very large data volume exists, which has a high degree of logical structure and many variables. Representative cases include point-of-sale transactions data and call-center records data.

Clustering can be performed to any domain, but it has been primarily applied on biomedicine, financial analysis, and for outlier analysis in general.

2.4 Training and Retraining with the Data Miner

Following the design primitives imposed by the abovementioned mechanism (Figure 5.7), the Data Miner has been developed as a GUI-based wizard that guides the user from the initial step of data loading, until the final step of creating the PMML document and embedding it into the reasoning mechanism of the agents. At first, the user launches the Data Miner and specifies whether he/she will train a new agent or retrain an existing one (Figure 5.8).

Figure 5.8. Launching Data Miner

In the former case, the Ontology panel is launched to allow the user to choose and load the ontology that will be used for the specific train-

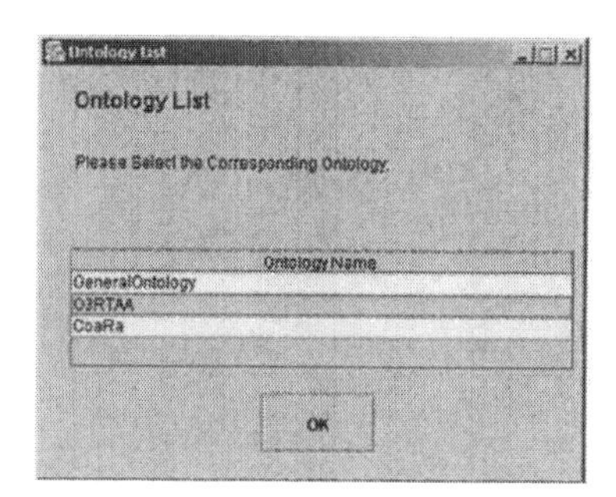

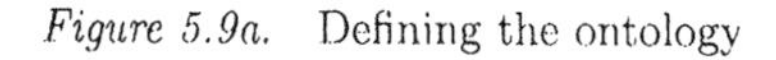
Figure 5.9a. Defining the ontology

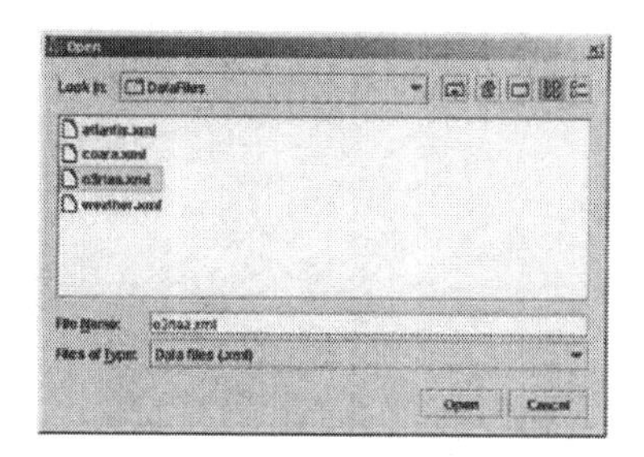

Figure 5.9b. Specifying the input file containing the training dataset

ing dataset (Figure 5.9a). In the latter case of retraining, instead of the ontology pane, the Decision Structure panel is launched, to enable selection of the KM that must be updated.

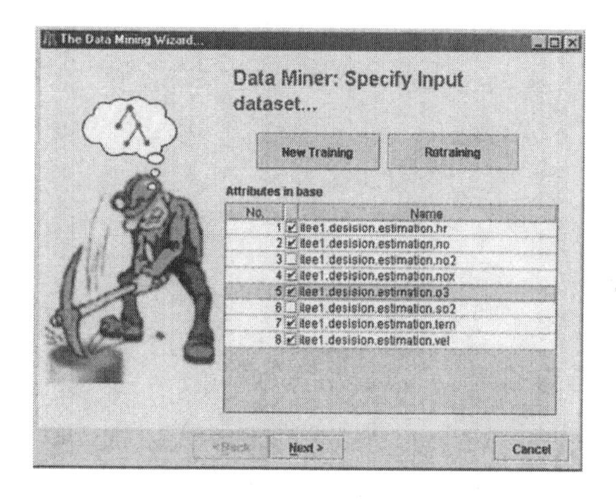

Figure 5.10. Preprocessing data

The corresponding XML data file is selected through the Selection Panel (Figure 5.9b), it is parsed, and then validated for syntactical correctness and ontology consistency. After that the data are loaded to the Data Miner and the user can decide on the attributes on which data mining will be performed, in case the dataset provided represents a superset of the desired one (Figure 5.10).

On the next step, the appropriate DM technique for the problem at hand and the most suitable algorithm is determined (Figure 5.11). For

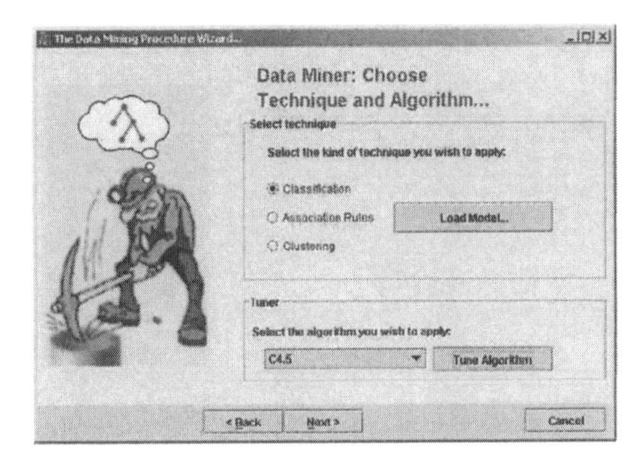

Figure 5.11. Selecting the proper DM technique and algorithm

each one of the algorithms selected, different fine tuning options are provided (Figure 5.12).

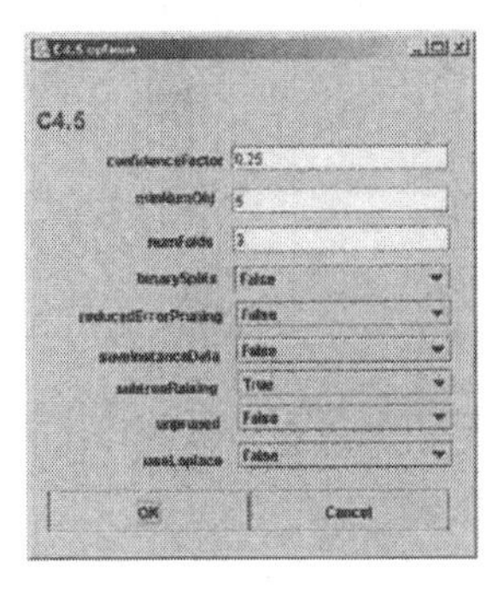

Figure 5.12. Tuning the selected algorithm

Training, testing and validation options are specified (Figure 5.13), i.e., whether training on the specified dataset is going to be performed or testing against another dataset, cross validation or percentage splitting – always with respect to the DM technique used.

The next step in the procedure involves setting the output options (Figure 5.14). The user must indicate whether to save or not the contents of the output buffer, the extracted model in Data Miner–compatible format, and/or PMML. The output buffer of the Data Miner holds information on evaluation metrics for each technique (mean square root errors, prediction accuracy, etc.). In addition, through this panel visualization capabilities are provided (only if the algorithm allows visualization).

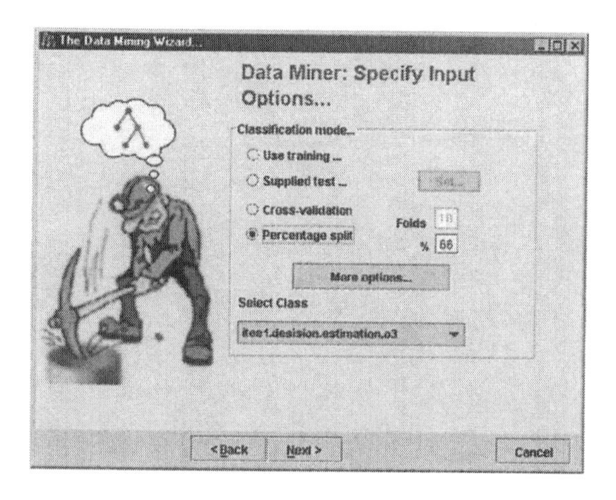

Figure 5.13. Specifying training parameters

An overview of the dataset options and the algorithm tuning, training and output defined parameters is provided in the last panel of the Data Miner, in order for the user to check for any errors or omissions. Then the data mining procedure is initiated. If the resulting decision model is deemed satisfactory by the user (Figure 5.15), then it is accepted and the corresponding PMML document is constructed.

This document is then parsed, translated into JESS blocks, and inserted into the agent's behavior, according to the mechanism described previously. Figure 5.16 summarizes the whole procedure.

Figure 5.14. Specifying output options

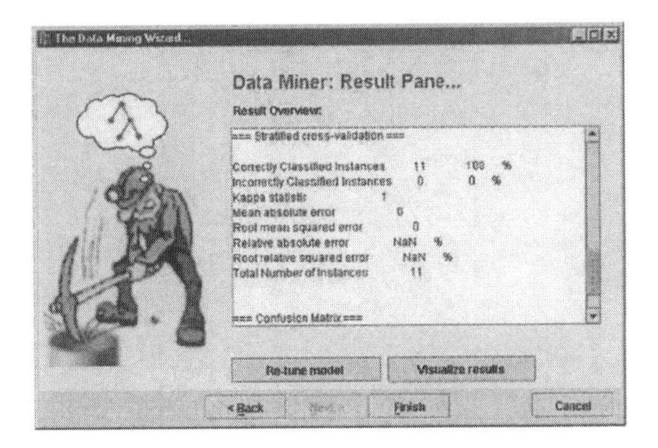

Figure 5.15. The outcome of the data mining process

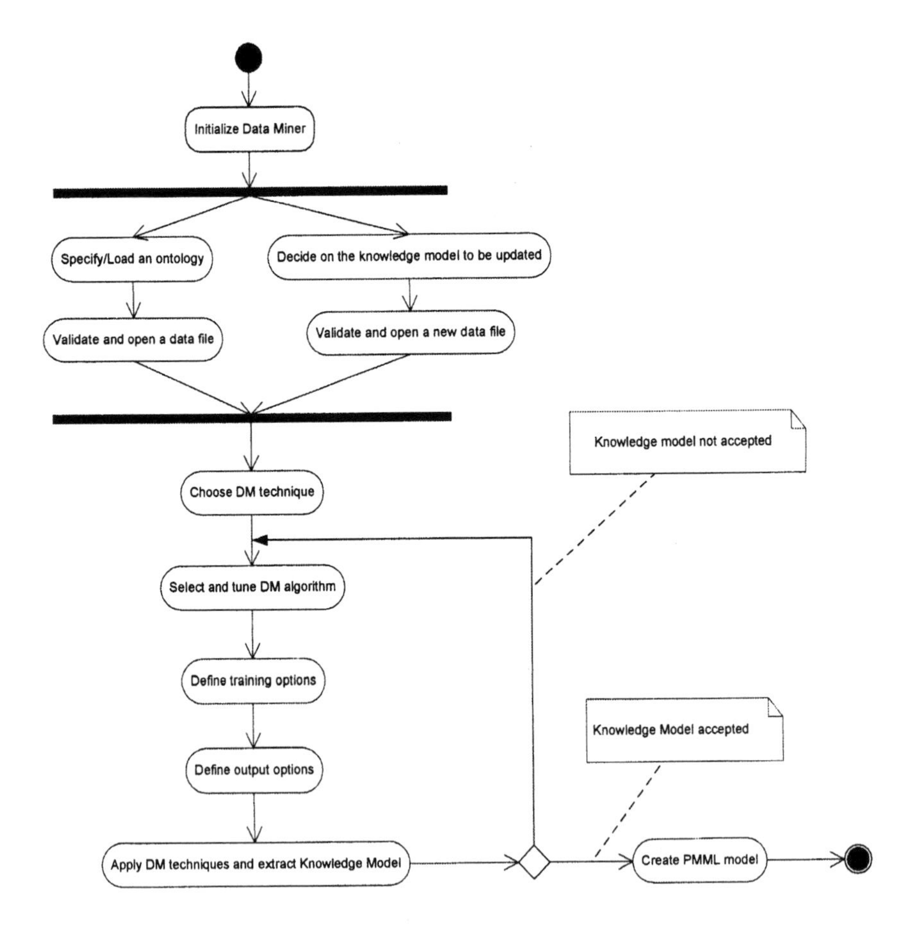

Figure 5.16. The functionality of Data Miner

PART III

KNOWLEDGE DIFFUSION: THREE REPRESENTATIVE TEST CASES

Chapter 6

DATA MINING ON THE APPLICATION LEVEL OF A MAS

In the first case of dynamic knowledge diffusion to MAS, knowledge is extracted by the application of DM techniques on historical application data. In such systems, an adequately large dataset is required (the bigger the better!), in order to produce a valid knowledge model, which reflects the trends in the data. A wide range of application domains may produce and maintain such large data volumes, including environmental systems, transaction systems, data warehouses, web logs etc. Special attention should be drawn, however, to Enterprise Resource Planning (ERP) systems, which are employed by all kinds of companies and for diverse purposes. Managing enterprise resources is an inherently distributed problem, that requires intelligent solutions. In addition, ERP systems generate and maintain large data volumes and DM techniques have already been exploited for improving several of the processes that are involved in the ERP process chain. These characteristics dovetail nicely with the prerequisites for dynamic knowledge diffusion and have led us to selecting the ERP application domain as the most suitable for demonstrating the methodology presented in the previous chapter.

1. Enterprise Resource Planning Systems

ERP systems are business management tools that automate and integrate all company facets, including real-time planning, manufacturing, sales, and marketing. These processes produce large amounts of enterprise data that are, in turn, used by managers and employees to handle all sorts of business tasks such as inventory control, order tracking, customer service, financing and human resources [Levi et al., 2000].

Despite the support current ERP systems provide on process coordination and data organization, most of them – especially legacy systems

– lack advanced Decision-Support (DS) capabilities, resulting therefore in decreased company competitiveness. In addition, from a functionality perspective, most ERP systems are limited to mere transactional IT systems, capable of acquiring, processing, and communicating raw or unsophisticated processed data on the company's past and present supply chain operations [Shapiro, 1999]. In order to optimize business processes in the tactical supply chain management level, the need for analytical IT systems that will work in close cooperation with the already installed ERP systems has already been identified, and DS-enabled systems stand out as the most successful gateway towards the development of more efficient and more profitable solutions. Probing even further, Davenport [Davenport, 2000] suggests that decision-making capabilities should act as an extension of the human ability to process knowledge and proposes the unification of knowledge management systems with the classical transaction-based systems, while Carlsson and Turban [Carlsson and Turban, 2002] claim that the integration of smart add-on modules to the already established ERP systems could make standard software more effective and productive for the end-users.

The benefits of incorporating such sophisticated DS-enabled systems inside the company's IT infrastructure are analyzed by Holsapple and Senna [Holsapple and Sena, 2004]. The most significant, among others, are:

1. Enhancement of the decision maker's ability to process knowledge.
2. Improvement of reliability of the decision support processes.
3. Provision of evidence in support of a decision.
4. Improvement or sustainability of organizational competitiveness.
5. Reduction of effort and time associated with decision-making, and
6. Augmentation of the decision makers' abilities to tackle large-scale, complex problems.

Within the context of Small and Medium sized Enterprises (SMEs) however, applying analytical and mathematical methods as the means for optimization of the supply chain management tasks is highly impractical, being both money– and time–consuming [Choy et al., 2002; Worley et al., 2002]. This is why alternative technologies, such as Data Mining and Agent Technology have already been employed, in order to provide efficient DS-enabled solutions. The increased flexibility of multi-agent applications, which provide multiple loci of control [Wooldridge, 1999] can lead to less development effort, while the cooperation primitives that

Agent Technology adopts point to MAS as the best choice for addressing complex tasks in systems that require synergy of multiple entities. Moreover, DM has repeatedly been used for Market Trend Analysis, User Segmentation, and Forecasting. Knowledge derived from the application of DM techniques on existing ERP historical data can provide managers with useful information, which may enhance their decision-making capabilities.

Going briefly through related work, we see that DM and MAS have been used separately for efficient enterprise management and decision support. Rygielski et. al. [Rygielsky et al., 2002] have exploited DM techniques for Customer Relationship Management (CRM), while Choy et. al. [Choy et al., 2002; Choy et al., 2003] have used a hybrid machine learning methodology for performing Supplier Relationship Management (SRM). On the other hand, MAS integrated with ERP systems have been used for production planning [Peng et al., 1999], and for the identification and maintenance of oversights and malfunctions inside the ERP systems [Kwon and Lee, 2001].

Elaborating on previous work, we have integrated AT and DM advantages into a versatile and adaptive multi-agent system that acts as an add-on to established ERP systems. Our approach employs Soft Computing, DM, Expert Systems, standard Supply Chain Management (SCM) and AT primitives, in order to provide intelligent recommendations on customer, supplier, and inventory issues. The system is designated to assist not only the managers of a company – "Managing by wire" approach [Haeckel and Nolan, 1994]–, but also the lower-level, distributed decision makers – "Cowboys" approach [Malone, 1998]. Our framework utilizes the vast amount of corporate data stored inside ERP systems to produce knowledge, by applying data mining techniques on them. The extracted knowledge is diffused to all interested parties via the multi-agent architecture, while domain knowledge and business rules are incorporated into the system by the use of rule-based agents. Our approach merges the already proven capabilities of data mining with the advantages of multi-agent systems in terms of autonomy and flexibility, and therefore promises a great likelihood of success.

2. The Generalized Framework

The generalized *Intelligent Recommendation Framework* (IRF) follows the methodology presented in the previous chapter. It consists of four discrete layers, as shown in Figure 6.1.

1) The *Communication* Layer, which is responsible for retrieving and storing data to the database of the ERP system,

2) The *Information Processing* Layer, where DM techniques are applied on ERP data and useful KMs are extracted,

3) The *Decision Support* Layer, which organizes the extracted KMs and incorporates company policy into a unified system recommendation and,

4) The *Graphical User Interface* Layer, which illustrates the final recommendation to the end user, either directly or indirectly.

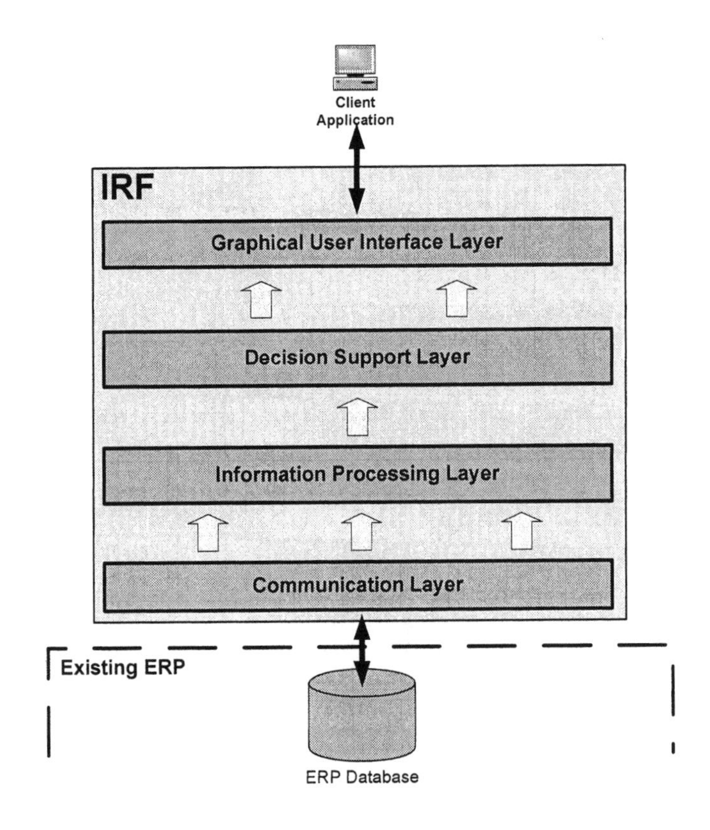

Figure 6.1. The layers of IRF

Due to the distributed nature of the IRF, the different layers are implemented with the use of software agents, which are responsible for all the core functions of the system. All inputs are transformed into ACL messages using the FIPA-SL0 language [The FIPA Foundations, 2000], since the MAS is FIPA-compliant [The FIPA Foundations, 2003].

In order for the reader to better comprehend the way information is diffused through the layers of IRF, we provide a sample use case:

The arrival of a new customer order triggers the IRF operation. First, all customer order preferences are gathered by the system operator via a front-end agent and are then transferred to the backbone (order) agents for processing. The order processing agents are of different types, each one related to a specific entity of the supply chain (company, customers, suppliers, products), and they manage entity-specific data. Communication with the ERP database is established via another agent. The profiles of all entities related to the recommendation process, which have been previously constructed using DM techniques, are recalled from the profile repository. These profiles, together with other order data, are used by the agents in the *Decision Support* layer to produce the recommendation. Finally, the front-end agent returns to the operator the intelligent recommendations, along with an explanatory memo. These recommendations are not meant to substitute the human operators, rather to assist them in their effort to efficiently manage customer orders and company supplies and, ultimately, to help the company realize a bigger increase profit.

2.1 IRF Architecture

The general IRF architecture is illustrated in Figure 6.2. The IRF agents belong to one of six different agent types ($Q_1 - Q_6$) and are listed in Table 6.1. The main characteristics and the functionality of each type are discussed in the following paragraphs.

Table 6.1. The IRF agent types and their functionality

Agent type	Name	Abbreviation	Functionality
Q_1	Customer Order Agent	COA	GUI agent
Q_2	Recommendation Agent	RA	Organization & Decision Making agent
Q_3	Customer Profile Identification Agent	CPIA	Knowledge Extraction agent
Q_4	Supplier Profile Identification Agent	SPIA	Knowledge Extraction agent
Q_5	Inventory Profile Identification Agent	IPIA	Knowledge Extraction agent
Q_6	Enterprise Resourse Planning Agent	ERPA	Interface agent

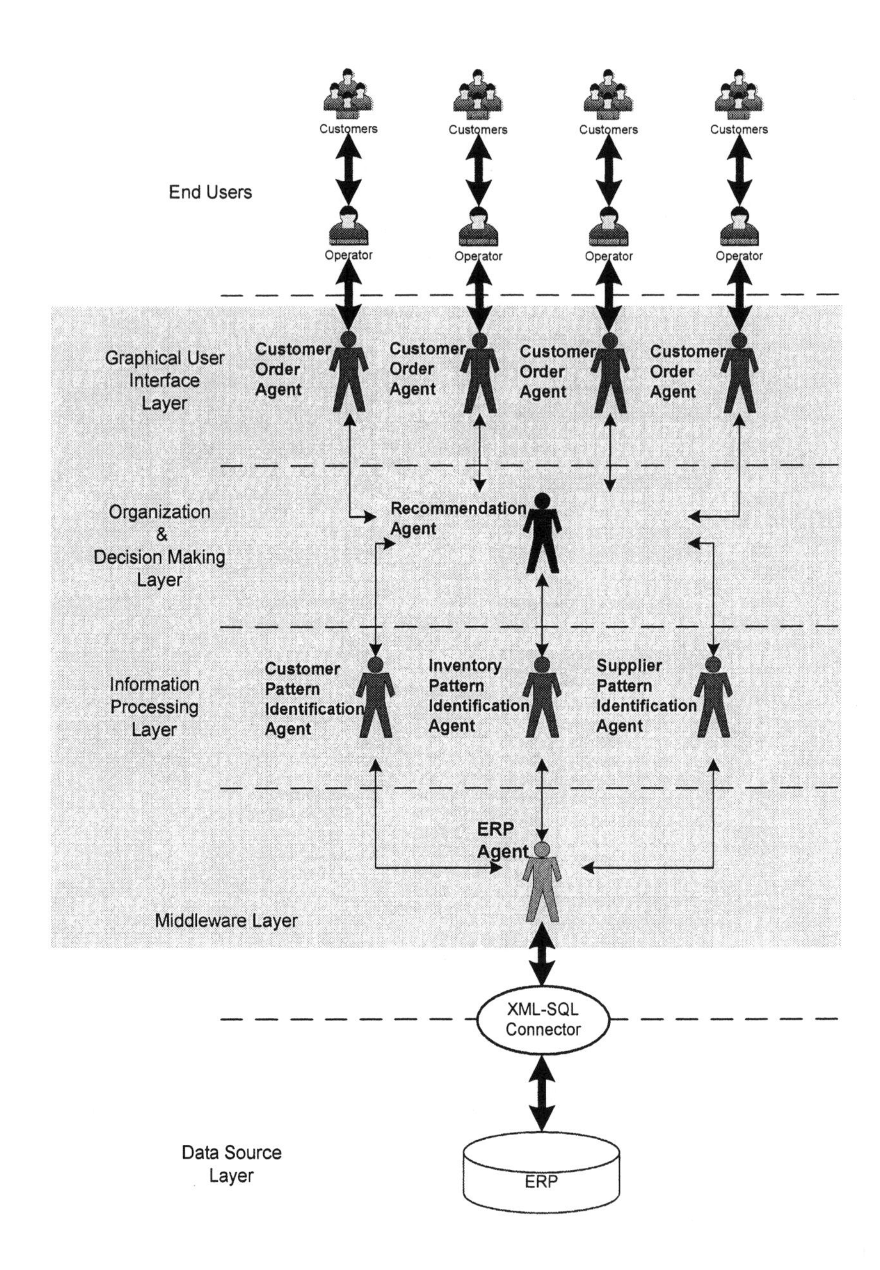

Figure 6.2. The IRF architectural diagram

2.1.1 Customer Order Agent type

COA is an interface agent that may operate at the distribution points, or at the telephone center of an enterprise. COA enables the system operator to: a) transfer information into and out of the system, b) input order details into the system, and c) justify, by means of visualization tools, the proposed recommendations. When an order arrives into the system, COA provides the human agent with basic functionalities for inserting information on the customer, the order details (products and their corresponding quantities), payment terms (cash, check, credit etc.), backorder policies and, finally, the party (client or company) responsible for transportation costs. COA also encompasses a unit that displays information in various forms to explain and justify the recommendations issued by the RA.

2.1.2 Recommendation Agent type

The RA is responsible for gathering the profiles of the entities involved in the current order and for issuing recommendations. By distributing the profile requests to the appropriate *Information Processing Layer* agents (CPIA, SPIA and IPIA - each one of them operating on its own control thread), and by exercising concurrency control, this agent diminishes the cycle-time of the recommendation process. RA is a rule-based agent implemented using the Java Expert System Shell (JESS) [Friedman-Hill, 1998]. Static and dynamic business rules can be incorporated into the RA. The latter must be written into a document that the agent can read during its execution phase. In this way, business rules can be modified on-the-fly, without the need of recompiling, or even restarting the application.

2.1.3 Customer Profile Identification Agent type

CPIA is designed to identify customer profiles, utilizing the historical data maintained in the ERP system. The process can be described as follows: Initially, managers and application developers produce a model for generating the profiles of customers. They select the appropriate customer attributes that can be mapped from the data residing in the ERP database; these are the attributes that are considered instrumental for reasoning on customer value. Then, they decide on the desired classification of customers, i.e., added-value to the company, discount due to past transactions etc. CPIA, by the use of clustering techniques, analyzes customer profiles periodically, and stores the outcome of this analysis into a profile repository for posterior retrieval. When a CPIA is asked to provide the profile of a customer, the current attributes of the specific customer are requested from the ERP database and are matched against those in the profile repository, resulting into the identification of the group the specific customer belongs to. During the development phase, one or more CPIA agents can be instantiated, and the distinction of CPIAs into training and recommendation ones, results to quicker response times when learning and inference procedures overlap.

2.1.4 Supplier Pattern Identification Agent type

SPIA is responsible for identifying supplier profiles according to the historical records found in the ERP database. In a similar to CPIA manner, managers identify the key attributes for determining a supplier's value to the company and their credibility. SPIA then generates supplier profiles and updates them periodically. For every requested item in the current order, the RA identifies one or more potential suppliers and requests their profiles from the SPIA. SPIA has to retrieve the current records of all the suppliers, identify for each one the best match in the profile repository, and return the corresponding profiles to the RA. Then RA can select the most appropriate supplier combination (according to its rule engine), and recommend it to the human operator. SPIA is also responsible for fetching to RA information about a specific supplier, such as statistical data on lead-times, quantities to be procured etc.

2.1.5 Inventory Profile Identification Agent type

IPIA is responsible for identifying product profiles. Product profiles comprise raw data from the ERP database (i.e., product price, related store, remaining quantities), unsophisticated processed data (for example statistical data on product demand) and intelligent recommendations on products (such as related products that the customer may be willing to purchase). Once more, managers and application developers have to identify the company priorities and map the profile to the data maintained by the ERP. Besides the directly–derived data, IPIA is responsible for identifying buying patterns. Market basket analysis can be performed with the help of association rule extraction techniques. Since this process is, in general, time-consuming, two or more IPIAs can be instantiated to separate the recommendation from the learning procedure.

2.1.6 Enterprise Resource Planning Agent type

ERPAs provide the middleware between the MAS application and the ERP system. These agents behave like transducers [Genesereth and Ketchpel, 1994], because they are responsible for transforming data from heterogeneous applications into message formats that agents can comprehend. An ERPA handles all queries posted by CPIAs, IPIAs, and SPIAs by connecting to the ERP database and fetching all the requested data. It works in close cooperation with an XML connector which relays

XML-SQL queries to the ERP and receives data in XML format. ERPA is the only IRF agent type that needs to be configured properly, in order to meet the connection requirements of different ERP systems.

2.2 Installation and Runtime Workflows

Once a company chooses to add IRF to its already operating ERP system, a few important steps have to be performed. The installation procedure of the IRF is shown in Figure 6.3.

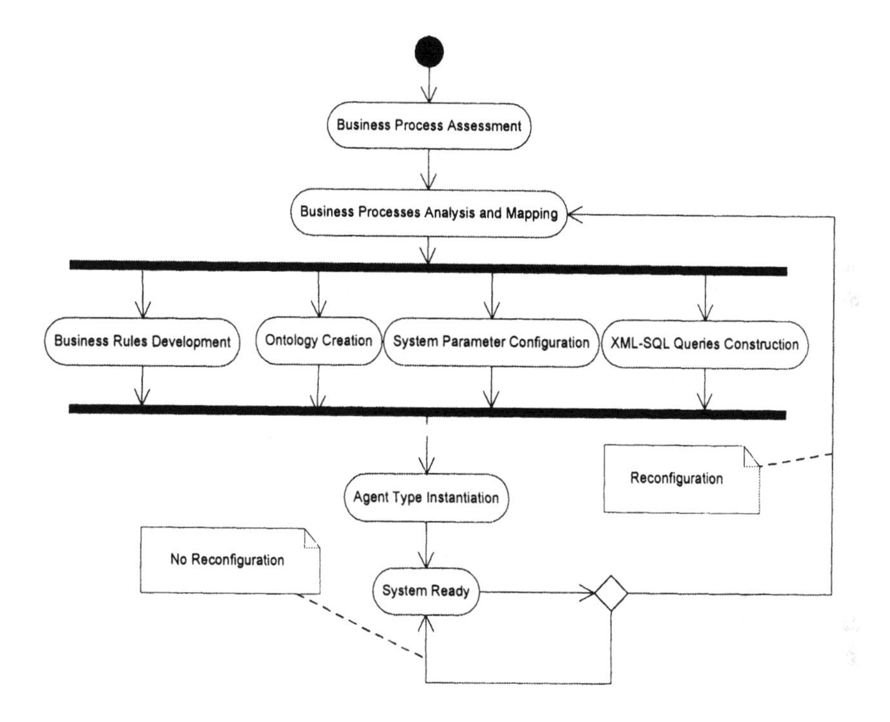

Figure 6.3. Installing IRF on top of an existing ERP

At first, the company's business process expert, along with the IRF application developers have to make a detailed analysis and assessment of the current customer order, inventory and products procurement processes. The results are mapped to the recommendation process of the add-on and the relevant datasets are delineated in the ERP.

After modeling the recommendation procedure according to the needs of the company, parallel activities for producing required documents and templates for the configuration of the MAS application follow. Fixed business rules incorporating company policy are transformed to expert system rules, XML-SQL queries are built and stored in the XML documents repository, ontologies (in RDFS format) are developed for the messages exchanged and for the decision on the workflow of the agents, agent types instantiation requirements are defined (at different workstations and cardinalities) and other additional parameters are configured (i.e., simple retraining time-thresholds, parameters for the data-mining

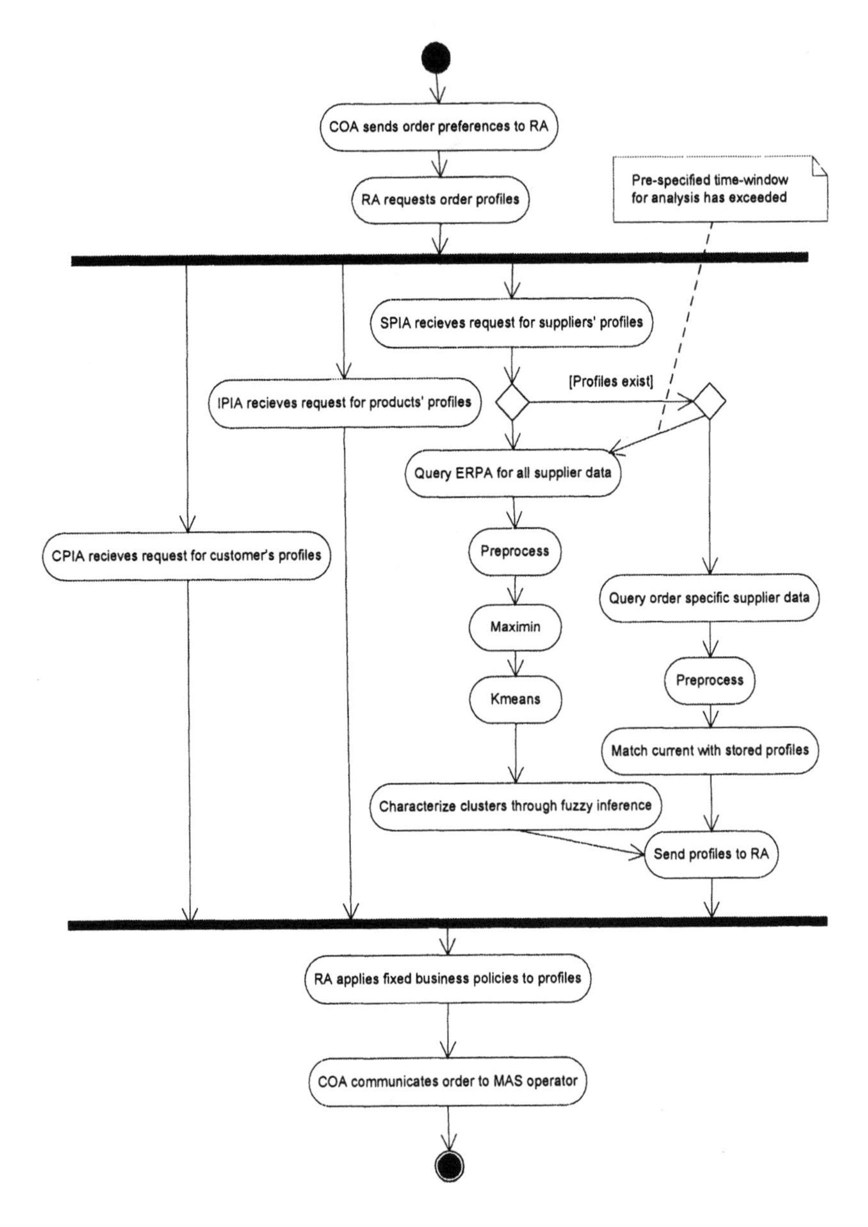

Figure 6.4. The Workflow of SPIA

algorithms, such as support and confidence for market basket analysis etc).

Once bootstrapped, reconfiguration of the system parameters is quite easy, since all related parameters are documents that can be conveniently reengineered. Figure 6.4 illustrates the workflow of the SPIA, where all the tasks described earlier in this section, can be detected. In case IRF needs to be modified due to a change in the company processes, the reconfiguration path must be traversed. The IPIA and CPIA workflows are similar and, thus, they are omitted.

2.3 System Intelligence

2.3.1 Benchmarking customer and suppliers

In order to perform customer and supplier segregation, CPIA and SPIA use a hybrid approach that combines data mining and soft computing methodologies. Clustering techniques and fuzzy inferencing are adopted, in order to decide on customer and supplier "quality". Initially, the human experts select the attributes on which the profile extraction procedures will be based on. These attributes can either be socio-demographic, managerial or financial data, deterministic or probabilistic. We represent the deterministic attributes, which are directly extracted from the ERP database by ERPA, as Det_i, $i = 1, ...n$, where n is the cardinality of the selected deterministic attributes. On the other hand, we represent the average (AVG) and standard deviation values (STD) of probabilistic variables, which are calculated by ERPA, as AVG_j and STD_j, $j = 1..m$, where m is the cardinality of the selected probabilistic attributes P_j.

Each customer/supplier is thus represented by a tuple:

$$< Det_1, ..., Det_n, AVG_1, STD_1, ..., AVG_m, STD_m) > \tag{6.1}$$

where $i = 1..n$, $j = 1..m$, $i + j > 0$.

Since real-world databases contain missing, unknown and erroneous data [Han and Kamber, 2001], ERPA preprocesses data prior to sending the corresponding datasets to the Information Processing Layer Agents. Typical preprocessing tasks are tuple omission and filling of missing values.

After the datasets have been preprocessed by ERPA, they are forwarded to CPIA and SPIA. Clustering is performed in order to separate customers/suppliers into distinct groups. The Maximin algorithm [Looney, 1997] is used to provide the number of the centers K that are formulated by the application of the K-means algorithm [McQueen, 1967]. This way K disjoint customer/supplier clusters are created.

In order to decide on customer/supplier clusters' added-value, CPIA and SPIA employ an Adaptive Fuzzy Logic Inference Engine (AFLIE), which characterizes the already created clusters with respect to an outcome defined by company managers, i.e., supplier credibility. Domain knowledge is incorporated into AFLIE [Freitas, 1999], providing to IRF the capability of characterization.

The attributes of the resulting clusters are the inputs to AFLIE and they may have positive ($\nearrow$) or negative ($\searrow$) preferred tendencies, depending on their beneficiary or harmful impact on company revenue. Once domain knowledge is introduced to AFLIE in the form of preferred tendencies and desired outputs, the attributes are fuzzified according to Table 6.2.

Table 6.2. Fuzzy variable definition and Interestingness of dataset attributes

Variable		Fuzzy Tuple
Input	**Preferred Tendency**	
Det_i	$\nearrow$	$< Det_i, [LOW, MEDIUM, HIGH], [Det_{i_1}, Det_{i_2}], Triangular >$
Det_i	$\searrow$	$< Det_i, [LOW, MEDIUM, HIGH], [Det_{i_1}, Det_{i_2}], Triangular >$
AVG_j	$\nearrow$	$< AVGj, [LOW, MEDIUM, HIGH], [AVG_{j_1}, AVG_{j_2}], Triangular >$
AVG_j	$\searrow$	$< AVGj, [LOW, MEDIUM, HIGH], [AVG_{j_1}, AVG_{j_2}], Triangular >$
Output	**Value Range**	
Y	Varies from Y_1 to Y_2 with a step of x	$< Y, [\#(Y_2 - Y_1)/x$ Incremental Fuzzy Values$], [Y_1, Y_2], Triangular >$

The probabilistic variables are handled in an adaptive way and are used as inputs only when Chebyshev's inequality (Eq. 6.2) is satisfied [Papoulis, 1984]:

$$P\{|P_j - AVG_j| \geq \varepsilon\} \leq \frac{(STD_j)^2}{\varepsilon^2}, \text{for any } \varepsilon > 0 \tag{6.2}$$

Eq. 6.2 ensures the concentration of probabilistic variables near their mean value, in the interval $(AVG_j - \varepsilon, AVG_j + \varepsilon)$. No attributes with high distribution are taken as inputs to the final inference procedure, avoiding therefore decision polarization.

The formulation of the inputs (3 values: $[LOW, MEDIUM, HIGH]$) leads to 3^{ν} *Fuzzy Rules* (FR), where ν is the number of AFLIE inputs. FRs are of type:

> **If** X_1 is $LX_1(k)$ and X_2 is $LX_2(k)$ and...and X_n is $LX_n(k)$
> **Then** Y is $LY(l)$, $k = 1..3$, $l = 1..q$,

where q is the cardinality of the fuzzy values of the output.

Triangular membership functions are adopted for all the inputs and outputs, whereas maximum defuzzification is used for crisping the FRs.

All inputs are assigned a *Corresponding Value* (CV), ranging from -1 to 1, according to their company benefit criterion (Table 6.2). The *Output Value* (OV) of Y is then calculated for each FR as:

$$OV = \sum_{i=1..n+m} w_i \cdot CV_i \tag{6.3}$$

where w_i is the weight of importance ($0 \leq w_i \leq 1$) of the ith input attribute.

The OVs are mapped to Fuzzy Values (FV), according to the degree of discrimination of the output decision variables. By categorizing the range of the output into q fuzzy values, the $OV \longrightarrow FV$ mapping is based on the following formula:

$$FV(OV) = RND\left[OV \cdot [\frac{2(n+m)}{q}]\right] \tag{6.4}$$

where $RND(x)$ is the rounding function of x to the closest integer (i.e., $MEDIUM$ for $x = 3$, $MEDIUM\_HIGH$ for $x = 4$ etc).

After all clusters have been characterized, the corresponding OVs, along with the cluster centers, are stored inside a profile repository for posterior retrieval. This process signals the end of the training phase of CPIA and SPIA.

In real time, when a new order comes into the system, RA requests the corresponding customer profile and the profiles of the suppliers that are related to the ordered products. CPIA and SPIA request, in turn, the attributes of these entities from ERPA, and match them against the profiles stored inside the profile repository, by the use of the Assigned Cluster (AC) criterion. AC is a closeness-to-cluster-center function, given by the following equation:

$$AC = \min_{i=1..k}\{\sqrt{\sum_{i=1}^{n+m}(c_i - xc_{ji})}\} \tag{6.5}$$

where k is the number of clusters, n the number of attributes, c_i is the ith attribute value of the cluster center vector $c = (c_1, c_2, ..., c_n)$, and xc_{ij} the ith attribute value of the jth current vector $xc_j = (xc_{j1}, xc_{j2}, ..., xc_{jn})$. The winning cluster along with its OV is returned to RA.

2.3.2 IPIA products profile

The IPIA plays a dual role in the system:

1. It fetches information on price, stock, statistical data about demand faced by the ordered products, and
2. It provides recommendations on additional items to buy, based on association rule extraction techniques.

In order to provide adaptive recommendations on ordering habits, IPIA incorporates knowledge extracted by the Apriori algorithm ([Amir et al., 1999; Ganti et al., 1999]. The association rules extracted are stored inside the profile repository for later retrieval.

Special attention should be drawn to the fact that the transactions included into the dataset to be mined may span several different customer order periods. XML-SQL queries can be adapted to perform data mining either to the whole dataset or the datasets of specific periods. Thus, IPIA is highly adaptable, both for companies in the general merchandize domain, but also for companies that sell seasonal goods (for example toys). The recommendations of IPIA, as well as the information concerning stock availability and price, are sent to the RA.

2.3.3 RA Intelligence

As mentioned earlier, RA is an expert agent that incorporates fixed business policies applied to customers, inventories, and suppliers. These rules are related, not only to raw data retrieved from the ERP database and order preferences provided by customers, but also to the extracted knowledge provided by the Information Processing agents. There are three distinct rule types that RA can realize:

1. Simple $< If \ldots Then \ldots >$ statements,
2. Rules describing mathematical formulas, and
3. Rules providing solutions to search problems and constraint satisfaction problems.

An example is provided below for each one of these rule types:

Example 1: Simple Rules

Additional discounts or burdens to the total price of an order can be implemented by the use of simple rules (knowledge extracted is denoted in bold):

a. IF ($TotalOrderRevenue >= 100$) AND (**CustomerValue** $= LOW$) THEN $TotalDiscount+ = 5\%$;

b. IF (**CustomerValue** $= LOW$) THEN $TotalDiscount- = 5\%$;

c. IF ($ProductType = ChristmasProducts$) AND ($TotalQuantity >= 100$) THEN **ProductDiscount+** $= 10\%$;

d. IF (**RecommendedProductsPurchased** $= True$) THEN $ProductDiscount+ = 5\%$;

Example 2: Mathematical Formulas

a) Re-order/Order-up-to-level metric sS

The re-order/order-up-to-level-point metric (sS) provides efficient inventory management for either no-fixed cost orders or fixed cost orders [Levi et al., 2000]. In the case of no-fixed cost orders (where $s = S$), the reorder point is calculated as:

$$sS = AVGD \cdot AVGL + z \cdot \sqrt{AVGL \cdot STDD^2 + AVGD^2 \cdot STDL^2} \quad (6.6)$$

where z is a constant chosen from statistical tables to ensure the satisfaction of a pre-specified value for the company's service level. Table 6.3 illustrates the value of z in correlation with the desired service level. In most legacy ERP systems such attributes have to be provided by users and cannot be derived automatically.

Table 6.3. Service Level and corresponding z Value

Service Level	90%	91%	92%	93%	94%	95%	96%	97%	98%	99%	99.9%
z	1.29	1.34	1.41	1.48	1.56	1.65	1.75	1.88	2.05	2.33	3.08

b) Splitting Policy

A splitting policy is applied when company stock availability cannot satisfy order needs. Upon arrival of a new order, the quantity of ordered

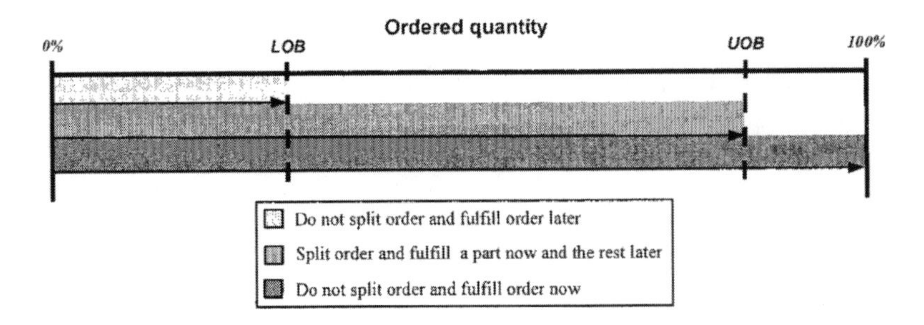

Figure 6.5. RA order splitting policy

items and available stock are cross-checked. If the requested quantities are available, the order is fulfilled immediately. Otherwise, the final supplying policy that the RA recommends is set according to the schema illustrated in Figure 6.5.

The LOB and UOB thresholds depend on the estimated customer value. In case we choose to incorporate product discount and customer priority into our splitting policy (for example, customers that enjoy better discount and have a higher priority to have a lower LOB and an higher UOB), we may adjust LOB and UOB according to the following equations:

$$LOB = \alpha_l \cdot \exp[-(b_{pl}\hat{p} + b_{dl}\hat{d})] \tag{6.7}$$

$$UOB = \alpha_u \cdot \exp(b_{pu}\hat{p} + b_{du}\hat{d}) \tag{6.8}$$

where $\hat{p}$ is the priority normalized factor, $\hat{d}$ is the discount normalized factor, while the weighting factors $< \alpha_l, b_{pl}, b_{dl}, \alpha_u, b_{pu}, b_{du} >$ are estimated in order to satisfy minimal requirements on LOB and UOB range.

If available stock is below $LOB\%$ of the ordered quantity, the entire order is put on hold until the company is supplied with adequate quantities of the ordered item. When item availability falls within the $[LOB - UOB]\%$ range of the ordered quantity, the order is split. All available stock is immediately delivered to the customer, whereas the rest is ordered from the appropriate suppliers. Finally, in case the available stock exceeds $UOB\%$ of the ordered quantity, the order is immediately preprocessed and the remaining order percentage is ignored.

Example 3: Problem Searching

a) Problems that require heuristics application and/or constraint satisfaction

Based on raw data from the ERP and on knowledge provided by SPIA, Recommendation Agents can yield solutions to problems like the selection of the most appropriate supplier with respect to their added-value, proximity to the depleted company store, or the identification and application of an established contract.

b) Enhanced Customer Relationship Management

Using the knowledge obtained by customer clustering, RA can implement a variety of targeted discount strategies in the form of crisp rules. Thus, the company has additional flexibility in its efforts to retain valuable customers and entice new ones with attractive offers [Rust et al., 2000].

3. An IRF Demonstrator

In order to demonstrate the efficiency of IRF, we have developed IPRA [Symeonidis et al., 2003], an Intelligent Recommender module that employs the methodology presented in Chapter 5. The system was integrated into the IT environment of a large retailer in the Greek market, hosting an ERP system with a sufficiently large data repository. IPRA was slightly customized to facilitate access to the existing Oracle^{TM} database.

Our system proved itself capable of managing over 25.000 transaction records, resulting in the extraction of truly "smart" suggestions. The CPIA and the SPIA performed clustering of over 8.000 customers (D_{IQ_3} dataset) and 500 suppliers (D_{IQ_4} dataset), respectively, while IPIA performed association rule extraction on 14125 customer transactions (D_{IQ_5} dataset).

All the attributes used by the Information Processing agents as inputs for DM, their corresponding preferred tendency, the inputs of the RA JESS engine, as well as the outputs of the IPRA system and their value range, are listed in Table 6.4.

The Information Processing agents of IPRA, in order to provide RA with valid customer and supplier clusters, as well as interesting additional order items, performed DM on the relevant datasets. For the specific company, CPIA and SPIA have identified each five major clusters representing an equal number of customer and supplier groups, respectively. Resulting customer (supplier) clusters, as well as the discount

Table 6.4. IPRA inputs and outputs

CPIA		SPIA		IPIA	RA
Input	Preferred Tendency	Input	Preferred Tendency	Input	Input
Account balance	↘	Account balance	↘	Stock Availability	Ordered Quantity
Credit Limit	↗	Credit Limit	↗	Item price	Stock Availability
Turnover	↗	Turnover	↗	Supplier ids	Re-order metric
Average Order Periodicity	↘	Average Order Completion	↘	Average Item Turnover (AIT) for the last two years	Supplier Geographic Location
Standard deviation of Order	-	Standard deviation of Order	-	Monthly Standard Deviation of	Lower Order Break-point
Average Order Income	↗	Average Payment Terms	↘		Upper Order Break-point
Standard deviation of Order Income	-	Standard deviation of Payment Terms	-		Customer Geographic Location
Average Payment Terms	↘	Supplier Geographic Location	↘		
Standard deviation of Payment Terms	-				
Customer Geographic Location	↘				
IPRA Outputs					
Output	Value Range	Output	Value Range	Output	Output
DISCOUNT	Varies from 0 – 30%, using a step of 5%	CREDIBILITY	Ranging from 0 – 1, using a step based on the number of supplier clusters	PROPOSED ORDER ITEMS	SPLITTING POLICY
PRIORITY	Varies from 0 – 3, using a step of 1				ADDITIONAL DISCOUNT
					CUSTOMER STATISTICS

and priority (credibility), calculated by the CPIA (SPIA) Fuzzy Inference Engine for each cluster, are illustrated in Table 6.5 and Table 6.6.

IPIA, on the other hand, has extracted a number of association rules from the records of previous orders, as shown in Table 6.7.

As already mentioned, upon receiving an order, the human agent collects all the necessary information, in order to provide IPRA with input. Data collected are handled by COA, the GUI agent of the system. An instance of the GUI is illustrated in Figure 6.6.

Table 6.5. The resulting customer clusters and the corresponding Discount and Priority values

Center ID	**Population (%)**	**Discount (%)**	**Priority**
0	0.002	20	High
1	10.150	10	Medium
2	46.600	15	Medium
3	22.240	10	Medium
4	20.830	5	Low

Table 6.6. The resulting supplier clusters and the corresponding Supplier Value towards the company

Center ID	**Population (%)**	**Value**
0	15.203	Low
1	10.112	Medium
2	25.646	Low
3	34.521	Medium
4	13.518	High

Table 6.7. The generated association rules with the predefined support and confidence thresholds.

Generated Rules	Support	Confidence
25	2%	90%
10	4%	90%

All information on items and quantities to be ordered, backorder policy, payment method, and transportation costs are given as input to IPRA. When the order process is initialized, COA forwards to the CPIA, SPIA, IPIA and RA respectively the already collected information. CPIA checks on the cluster the client falls into, SPIA decides on the best supplier, (according to his/her added-value), in case an order has to be placed to satisfy customer demand, IPIA proposes additional items for the customer to order, and all these decisions are passed on to the RA, which decides on the splitting policy, (if needed) and on additional discount.

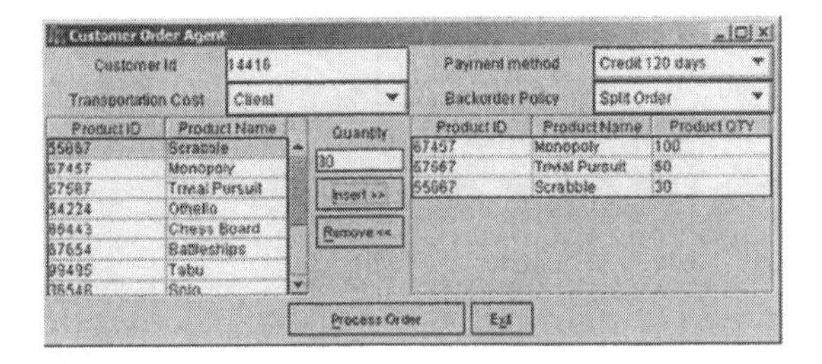

Figure 6.6. GUI of Customer Order Agent with information on the new order

Figure 6.7 illustrates the final recommendation created. Detailed information on the order and its products, customer suggested priority and discount, customer clusters, supplier suggested value and supplier clusters, additional order items, suggested order policy and statistics, are at the disposal of the human agent, to evaluate and realize the transaction at the maximal benefit of the company.

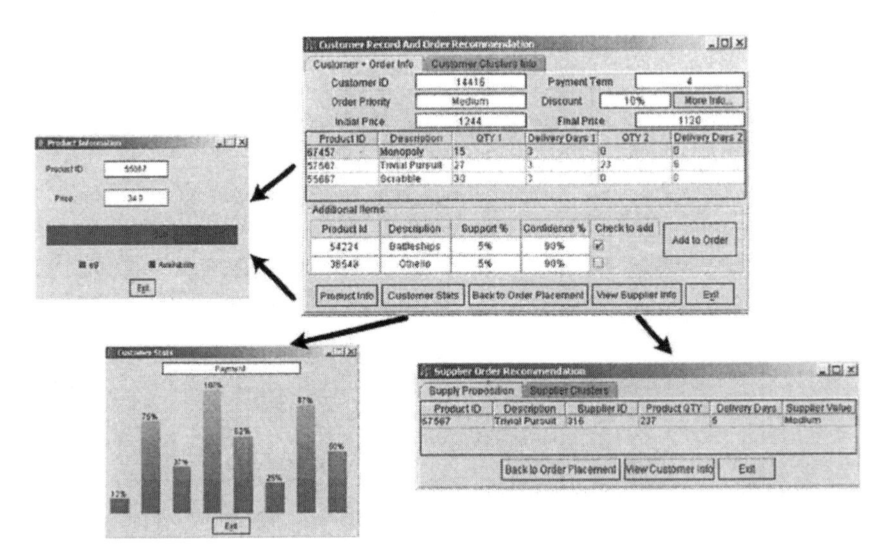

Figure 6.7. The final IPRA Recommendation

4. Conclusions

An ERP system, although indispensable, constitutes a costly investment and the process of updating business rules or adding customization modules to it is often unaffordable, especially for SMEs. The IRF methodology aspires to overcome the already mentioned deficiencies of non DS-enabled ERP systems, in a low-cost yet efficient manner. Knowledge residing in a company's ERP can be identified and dynamically incorporated into versatile and adaptable CRM/SRM solutions. IRF integrates a number of enhancements into a convenient package and establishes an expedient vehicle for providing intelligent recommendations to incoming customer orders and requests for quotes. Recommendations are independently and perpetually adapted, without an adverse impact on IRF run-time performance. IRF architecture ensures reusability and re-configurability, with respect to the underlying ERP. Table 6.8 summarizes the key enhancements provided by the augmentation of ERP systems with the IRF module.

Table 6.8. IRF enhancements to ERPs

	IRF + ERP	Legacy ERPs
Static Business Rules	**Yes** Provided as rule documents changed on the fly.	**Yes** Hard-coded by the ERP vendor.
Dynamic Business Rules	Applied to data + knowledge	Applied only to data
Market Basket Analysis	**Yes** Added online to the recommendation procedure	**No** (Unless external MBA is performed)
Recommendation Procedure	Automatically generated	Through reports
Inventory Management	Thresholds automatically adapted	Thresholds inserted manually if applicable (Unless SCM module incorporated)
Decision cycle-time	**Short** (Not related to database size)	**Long** (Related to database size)
Distributed Decision Making	**Yes** Recommendations can be used by lower level personnel	No
Adaptability	**High**	**Low**
Autonomy	**Yes**	**No**
Customers Intelligent Evaluation	**Yes**	**No** (Unless CRM module incorporated)
Suppliers Intelligent Evaluation	**Yes**	**No** (Unless SRM module incorporated)
Information Overload Reduction	**High**	**Small** (Through reports)
Cost of enhancement	**Low** (Use of AA platform)	**High** (Customization/third party DS COTS)

Chapter 7

MINING AGENT BEHAVIORS

The second level of dynamic knowledge infusion to MAS requires the existence of historical data describing the actions of agents within the MAS they reside. Such systems are mainly focused on the prediction of agent behaviors, and special attention is given on the way agent actions are modeled and represented. The core objective of agent action-based personalization is the discovery of a *recommendation set*, that will better predict the behavior of the agent "in action". In order for the reader to better comprehend the notions of agent behavior modeling and prediction, we have developed an agent-based framework (a recommendation engine for navigating the web), where agent actions are directly mapped to web user traversal decisions. The main issues related to the design and development of such a framework for predicting agent actions are discussed in this chapter, while the basic concessions made to enable agent cooperation are outlined. We also present κ-Profile, a new data mining mechanism for discovering action profiles and for providing recommendations on agent actions. Finally, the developed demonstrator is described and indicative experimental results are apposed.

1. Predicting Agent Behavior

1.1 The Prediction Mechanism

Let $P = \{p_1, p_2, \ldots, p_n\}$ be a set of possible actions that an agent may take during its execution phase. Parameter n varies from one operation cycle[1] to another. Let us also consider a set of m agent action bundles,

[1]We define an operation cycle as the sequence of actions performed by an agent in its effort to accomplish the specific task it has been assigned to

$B = \{b_1, b_2, \ldots, b_m\}$, where each action bundle $b_i \in B$ is a subset of P, and describes the actions taken by an agent throughout one operation cycle. Since all agent action bundles are defined on the action space P, each vector b has a length n and is of the form:

$$b = < w(p_1, b), w(p_2, b), \ldots, w(p_n, b) > \tag{7.1}$$

where $w(p_i, b)$ is a weight associated with action $p_i \in P$ and $0 \leq w(p_i, b) \leq 1$.

Let us consider a vector b that describes the actions taken by an agent within one operation cycle. The jth component $w(p_j, b)$ of this vector will be 0, if the agent has never taken action p_j, while $w(p_j, b) \neq 0$ when the agent has taken action p_j. In fact, the value of $w(p_j, b)$ increases if the action under consideration is of great importance. Let us now consider an operation cycle that has not yet terminated. Its corresponding vector has, in general, more null elements than a vector of an already terminated cycle. The scope of the prediction mechanism is to determine the vector components (i.e., the agent actions) that have a high probability of occurrence in the remainder of an ongoing operation cycle. This capability will effectively narrow the options of an agent to the most suitable ones and, thus, reduce the time required to compute its next step. The weight calculated for each action denotes the probability of its appearance in the corresponding operation cycle.

A mechanism for discovering common action patterns, which we call κ-Profile is employed to extract the recommendations. The κ-Profile involves a sequence of steps shown in Figure 7.1. It utilizes the Maximin [Looney, 1997] and K-Means [McQueen, 1967] algorithms, in order to cluster the set of vectors B, which essentially form the dataset D_{BT}, as defined in Chapter 5.

After clustering has been performed, a set of action bundle clusters, $BC = \{c_1, c_2, \ldots, c_k\}$ is extracted, where each c_i is a subset of B. For each action bundle cluster $c_i \in BC$, a representative vector is defined as the **cluster profile** cp_i. Vector cp_i is the vector closer to the cluster center. The components of cp_i that fall below a certain threshold μ are nullified. This way, only the most representative of the events in the cluster are taken into account. In the case of cluster c_i, for example, the profile cp_i is the set of $< action - weight >$ pairs, as defined by Eq. 7.2:

$$cp_i = \{< p, weight(p, cp_i) > \mid p \in P \text{ and } weight(p, cp_i) \geq \mu\} \tag{7.2}$$

The $weight(p, cp_i)$ of action p within profile cp_i is calculated using Eq. 7.3:

$$weight(p, cp_i) = \frac{1}{|c_i|} \sum_{b \in c_i} w(p, b) \tag{7.3}$$

where $w(p, b)$ is the weight of action p in action bundle $b \in c_i$. It is, therefore, obvious that each profile can be also represented as a vector of length n. Considering that the same representation mechanism is employed for the ongoing operation cycle vector, both profiles and action bundles can be manipulated as n-dimensional vectors in the action space.

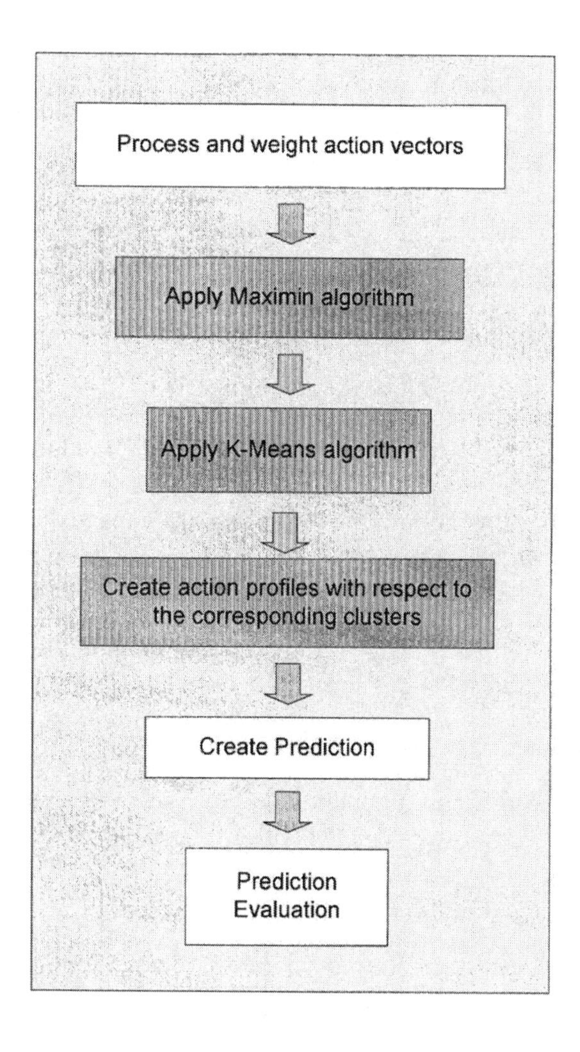

Figure 7.1. The κ-Profile mechanism

For example, a profile C, can be represented as:

$$C = \{w_1^C, w_2^C, ..., w_n^C\} \tag{7.4}$$

where

$$w_i^C = \begin{cases} weight(p_i, C), & p_i \in C \\ 0, & otherwise \end{cases} \tag{7.5}$$

Similarly, the ongoing operation cycle can be represented as a vector $S =< s_1, s_2, ..., s_n >$, where s_i is a weight denoting the significance of action p_i in the current cycle. The weighted values are calculated in the same manner as the weights for action vectors b are calculated (e.g. $s_i = 0$ if the agent has not taken action p_i). In both cases, a *Fuzzy Inference System* (FIS) is employed to produce the weights, which enables the incorporation of domain understanding into the prediction mechanism. The comparison of a certain profile with the vector representing the ongoing operation cycle is performed using the cosine coefficient, a metric widely used in information retrieval problems. $match(S, C)$, defined in Eq. 7.6, calculates the cosine of the angle of the two vectors S and C, by normalizing their dot product with respect to their moduli.

$$match(S, C) = \frac{\sum_k w_k^C \cdot s_k}{\sqrt{\sum_k (s_k)^2 \times \sum_k (w_k^C)^2}} \tag{7.6}$$

The actions that the prediction system recommends are determined through a *recommendation score*, defined for each action p_i in each of the already calculated profile vectors. This score is dependent on two factors:

1. The overall similarity of the current vector to the profile, and

2. The average weight of each action p_i in the profile

Given a profile C and an operation cycle vector S, the recommendation score $Rec(S, p_i)$ is calculated for each action p_i, according to Eq. 7.7:

$$Rec(S, p_i) = \sqrt{weight(p_i, C) \cdot match(S, C)} \tag{7.7}$$

Finally, a Next Action Recommendation score, $NAR(S)$, is compiled for the current action bundle S, containing only actions with recommendation scores that exceed a certain threshold, ρ, for all profiles. That is:

$$NAR(S) = \{p_i \mid \mathrm{Rec}(s, w_i^C) \geq \rho\} \tag{7.8}$$

For each action appearing in more than one vectors, the maximum recommendation score is selected, from the corresponding profile. This way optimal coverage is achieved. Table 7.1 illustrates an example of a recommendation on action vector S, based on profile C. Recommendation is produced only on vector components that have a null value on S and a non-null value on C. For action p_2, for example, a recommendation score on S is calculated, according to Eq. 7.7. In this case, $Rec(S,p_2) = \sqrt{0.375 \cdot match(S,C)}$, where $match(S,C) = 0.52$. Finally, $Rec(S,p_2) = 0.442$.

Table 7.1. Recommending the next action

Actions P	Profile Vector C	Vector S	$Rec(S,p_i)$
p_1	0	0	0
p_2	0.375	0	0.442
p_3	0	0	0
p_4	0.3	0.6	0.395
p_5	0	0.4	0

1.2 Applying κ-Profile on MAS

The κ-Profile mechanism aims to predict future agent actions within an operation cycle, based on knowledge of prior actions of this and/or similar agents. For the ongoing operation cycle, a z-size window is employed. That is, only the last z actions of the agent can influence the outcome of the recommendation process. κ-Profile can be easily adapted to predict agent behavior, mapping the vector elements to agent actions.

Let us consider an agent authorized to execute a number of functions on files, as determined by its action space $P = \{Select, Open, Modify, Save, Close\}$. Binary weights are assigned to the elements of the action vectors. Let us consider a profile $C = \{Open, Modify, Save\}$. The corresponding profile vector is, thus, $cp_C = [0\ 1\ 1\ 1\ 0]$. In case actions $\{Open\}$ and $\{Modify\}$ are executed during the current cycle and the prediction window has been set to $z = 2$, action $\{Save\}$ will be recommended, according to profile C (Table 7.2).

Although excluding the two actions $\{Select\}$ and $\{Close\}$ from a set of five possible actions does not seem interesting, an equivalent reduction of the candidate space in a system with a large number $(n \geq 100)$ of options would be rather significant. κ-Profile provides this pruning mechanism through the grouping of action bundles into clusters and the identification of actions that are likely to occur next. These actions are

Table 7.2. An example on predicting the next action

Action	Profile	Recent History
Select	0	0
Open	1	1
Change	1	1
Save	1	0
Close	0	0

determined by their degree of participation into the profile(s) taken into account, and by the similarity of the ongoing operation cycle vector to it(them).

In our example, the participation degree of action $\{Save\}$ is 1. This action is proposed to be the next action for the ongoing operation cycle vector with an $1\times\{vector\ similarity\}$ similarity measure. The similarity measure is always < 1, since the current action bundle is generally different from the profile used for prediction. It should be denoted that the quality of the prediction is also related to the size of the historical dataset. A bigger dataset offers more training options, often leading to more accurate predictions.

In a sense, κ-Profile produces a type of association rules. For the current example, the rule would be:

$$\{Open\} \wedge \{Change\} \longrightarrow \{Save\} \tag{7.9}$$

Nevertheless, in the case of association rules, their participation degree (confidence) could be equal to 1, expressing certainty on the occurrence of the action.

The representation mechanism in our example is quite simple, since the weighted values are either 0s or 1s (the action has not/has been taken). In general, the mechanism is much more complicated and several parameters need to be considered (i.e., action frequency, action timing, etc.). The major advantage of κ-Profile is that it can manipulate vectors with fuzzy constituent values (fuzzy degrees of participation). In that case, the representation mechanism assigns values within the specified fuzzy interval, a flexibility that makes the κ-Profile mechanism suitable for a wide area of applications.

Important issues that deserve further study include the agent action model and the specification of the operation cycle goal. These issues are discussed next.

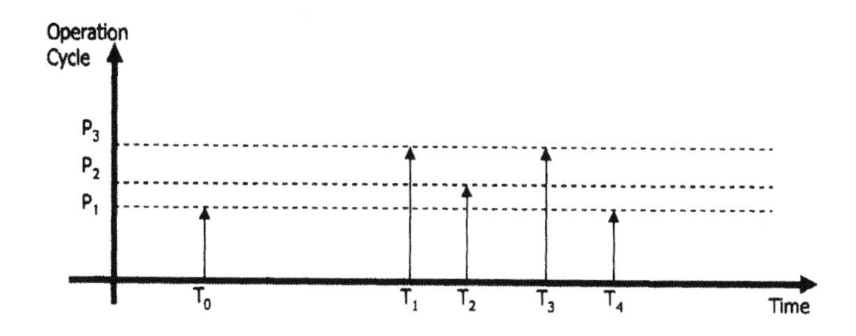

Figure 7.2. The evolution of an operation cycle

1.3 Modeling Agent Actions in an Operation Cycle

Let us consider some multi-agent application and let $P = \{p_1, p_2, \ldots, p_n\}$ be a finite set of possible agent actions. In general, agent actions are asynchronous events occurring at times T_i. The time interval between two actions is, in most cases, of great importance. In the case of a web application, for example, the time interval between two successive page visits is the time the user spent on the first site (possibly exposed to electronic advertising). In order to monitor an agent operation cycle we use an ordered, variable-length vector Π, whose elements are pairs of the form $< action, executiontime >$. The time intervals between consecutive actions can be easily calculated. In fact, proper processing of Π can produce useful information for the operation cycle.

As an example, let us consider a system where the set of possible actions is $P = \{p_1, p_2, p_3\}$. Figure 7.2 is a representation of an operation cycle with five agent actions occurring at times T_0 to T_4. Table 7.3 shows vector Π for this example.

Table 7.3. A vector representing the operation cycle

Vector Π
$< p_1, T_0 >$
$< p_3, T_1 >$
$< p_2, T_2 >$
$< p_3, T_3 >$
$< p_1, T_4 >$

According to the previous analysis, two things have to be determined, as far as the operation goal is concerned: a) the goal itself and, b) a terminating condition for the operation cycle. For an internet-based

MAS, the operation goal would be the transition of an agent to one of the available web pages of the site (each transition is considered to be an action). In this case, the prediction mechanism can be quite straightforward, because the operation cycle itself has to be predicted and no terminating condition needs to be specified. During the online process, only an action window (z) is needed for the system to predict. This is not the case, though, for the offline process, where operation cycle vectors must be created and stored. Since the profile vectors must be of equal length, a terminating condition is needed. This condition can be related to either time or change of application status, or even both. Thorough analysis has led to the decision that the optimal choice in such systems with "predicting" capabilities would be the definition of an agent action signaling the end of the current operation cycle. In the web traversing example, the terminating action could be the end of the user's web navigation.

1.4 Mapping Agent Actions to Vectors

Another important issue for the prediction mechanism is the transformation of the operation cycles to agent action vectors. κ-Profile mechanism requires equal-size vectors, with their elements (weights) valued in the [0,1] interval (Table 7.4). Binary weights could be assigned in the simplest of cases, merely denoting the execution or not of an action in a specific operation cycle. This approach is not suitable, however, for the cases when the same action is executed more than once within the same operation cycle.

It should be emphasized that the representation mechanism is closely related to the scope of the specific MAS. In a MAS focused on the analysis of consumer behavior, for example, the items and their quantities purchased within the same transaction would be of interest. In other types of applications, the time interval between actions is important (e.g., the browsing duration of a specific web page). It is, therefore, ob-

Table 7.4. Mapping agent actions to vectors

Vector π		Weighted Vector
$< p_1, T_0 >$		β_1
$< p_3, T_1 >$	$\longrightarrow$	β_2
$< p_2, T_2 >$		β_3
$< p_3, T_3 >$		β_4

vious that thorough analysis is required in order to achieve the proper modeling of the problem at hand.

The κ-Profile has been designed to recommend only actions of high importance. The mechanism perceives the importance of an action within the operation cycle through its weight (the corresponding element's weight). The action weights are calculated by the use of a FIS, that deals successfully with the issue of incorporating domain understanding. During fuzzification, the related parameters have to be defined and tuned, while the fuzzy rule base has to be created.

If the prediction system is developed following the methodology presented in Chapter 5, the only parameters that have to be defined are:

- The finite set of actions P
- The goal of the operation cycle and the terminating condition, and
- The parameters of the FIS, and their influence on the MAS performance.

1.5 Evaluating Efficiency

The efficiency of improving agent intelligence through data mining on agent behavior data can be measured at two levels:

a) the profile level, and

b) the prediction level

The behavior profiles extracted by the κ-Profile have to undoubtedly represent related behaviors among agents acting within the same application. If the profiles are successful, then **information personalization** can be achieved, which is of great importance to the performance of the system.

1.5.1 Profile efficiency evaluation

In order to evaluate profile efficiency, we can follow the analysis proposed by Perkowitz and Etzioni ([Perkowitz and Etzioni, 1998]). According to this analysis, an agent that has taken an action of the profile, will also take another action of the same profile, within the ongoing operation cycle. That is, if B is the set of profile vectors and cp_i is a profile, then B_{cp_i} is a subset of B, whose elements (b_j) contain at least one of the actions in cp_i. First, we must calculate the *average visit percentage* (AVP) for extracted profiles. The AVP of profile cp_i with respect to all the profile vectors is in this case given by Eq. 7.10:

$$AVP = \sum_{b \in B_{cp_i}} \frac{(\overrightarrow{b} \cdot \overrightarrow{cp_i})}{|b|} \tag{7.10}$$

The *weighted average visit percentage*, $WAVP$ metric is then calculated by dividing AVP with the sum of all profile cp_i elements' weights (Eq. 7.11).

$$WAVP = \frac{\left(\sum\limits_{b \in B_{cpi}} \frac{\overrightarrow{b} \cdot \overrightarrow{cp_i}}{|b|} \right)}{\left(\sum_{p \in cp_i} weight(p, cp_i) \right)} \tag{7.11}$$

Values of $WAVP$ closer to 1 indicate better profile efficiency.

1.5.2 Prediction system efficiency evaluation

Several different approaches can be followed, in order to evaluate the efficiency of a prediction system. Nevertheless, the primitives of a MAS impose an approach that is somewhat different to the ones used for estimation and classification systems. A recommendation engine, like the one described, does not require the rigidity of such systems. More than one suggestions, equally important, can be produced. Therefore, the prediction itself cannot be considered as a valid evaluation metric, and this is the reason for not employing the classic evaluation metrics, *precision* and *coverage*. Instead, the following approach has been adopted:

Let $AS = s_1, s_2, ..., s_n$ be the test set, i.e. a set of action vectors that were not used during the application of the prediction mechanism on the initial data (training set). Let R be the set of actions proposed by the prediction system for action s. If $R \cap s \neq 0$, then we consider the recommendation to be a success. The metric for evaluating the efficiency of the recommendation system is *PSSR* (Prediction System Success Rate), which is defined as the percentage of successful recommendations ($SRec$) to the sum of recommendations made ($ORec$):

$$PSSR = \frac{SRec}{ORec} \tag{7.12}$$

Values of $PSSR$ closer to 1 indicate better prediction efficiency.

2. A Demonstrator: Recommendation Engine for a Web Browser

In order to demonstrate the prediction mechanism and the diffusion of knowledge extracted by the use of DM techniques on agent behavior data, we present a MAS prototype, which realizes a recommendation engine for a large web site. Web browsing applications can employ software agents to monitor user actions and to possibly recommend products or other actions, following the analysis of user preferences. In our demonstrator, the action is defined as the transition to one of the available pages in the web site. In the following sections, the parameters of the system are specified and some typical results from its operation are illustrated.

2.1 System Parameters

It should be denoted that the developed application is focused on the way a user navigates through a specific web site, and not the entire web. Each one of the web site pages was mapped to an agent action. The web site used has 142 discrete web pages, i.e., 142 possible agent actions. Thus, $P = \{p_1, p_2, \ldots, p_{142}\}$.

The action bundle set (set of operation cycles) b is formulated according to Eq. 7.1, where the weights of the vector elements are specified with the use of a fuzzy inference system.

Two of the system parameters were considered important for a qualitative differentiation between user traversals:

i. The *time* a user spends on a certain web page

ii. The *frequency* at which a user visits a web page during an operation cycle

These two parameters were used as inputs to the recommendation engine FIS and are described next.

Table 7.5. Fuzzy variable *time* values and corresponding range

Fuzzy Values	Variable Range
VL	0-19 sec
L	10-50 sec
ML	20-120 sec
M	70-90 sec
H	130-∞ sec

2.1.1 The fuzzy variable *Time*

The *time* variable indicates the duration of the user visit on each web page of the site. Based on *time* one can speculate whether the user has found the contents of the web page interesting (spent reasonable time), trivial (spent almost no time), or if the user has neglected to terminate the web browser (the time interval was too long).

Table 7.5 lists the fuzzy values assigned to *time*, while Figure 7.3 illustrates the corresponding membership functions.

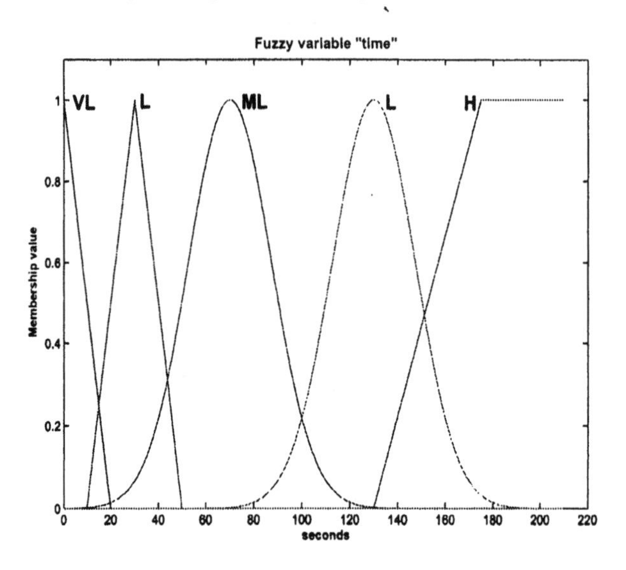

Figure 7.3. Fuzzy variable *time* values and corresponding membership functions

2.1.2 The fuzzy variable *Frequency*

The *frequency* variable indicates the number of visits a user pays to a certain web page within the same operation cycle. Increased page frequency implies increased page importance, unless the page is a navigational node. In such a case, the developed recommendation engine can easily bypass the page and assist users to reach their target faster and in a more efficient manner.

Table 7.6. Fuzzy variable *frequency* values and corresponding range

Fuzzy Values	Variable Range
Normal	0-1.7
Med	1-3.4
High	2.4-∞

The fuzzy values of *frequency* are specified in Table 7.6, while Figure 7.4 illustrates the corresponding membership functions.

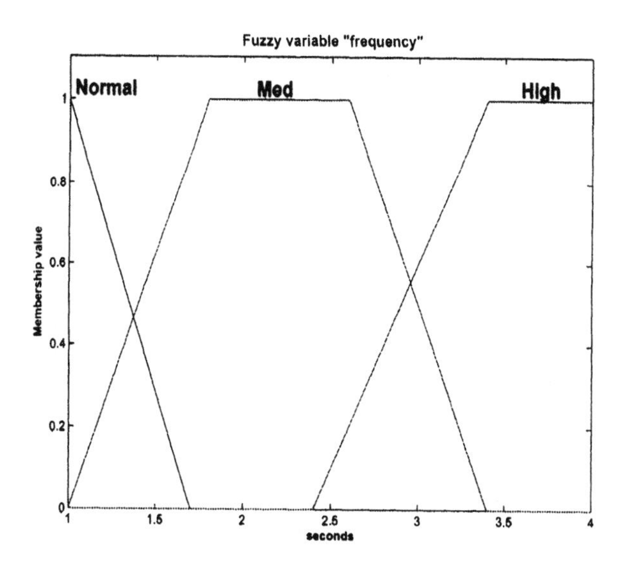

Figure 7.4. Fuzzy variable *frequency* values and corresponding membership functions

2.1.3 The output fuzzy variable *Weight*

The fuzzy variable *weight* is the output of the FIS and it is used to specify the weight of each action, with respect to *time* and *frequency*, within each one of the operation cycle vectors $b \in B$.

The fuzzy values of the output variable *weight*, along with their range are apposed in Table 7.7, while the corresponding membership functions are illustrated in Figure 7.5.

Table 7.7. Output variable *weight* fuzzy values and corresponding range

Fuzzy Values	Variable Range
VL	0-0.17
L	0.07-0.29
ML	0.18-0.45
M	0.31-0.69
MH	0.48-0.8
H	0.675-1

2.2 The Rules of the FIS

The FIS underlying concept is that it can incorporate domain understanding within the processes of a certain application. *Input-Output*

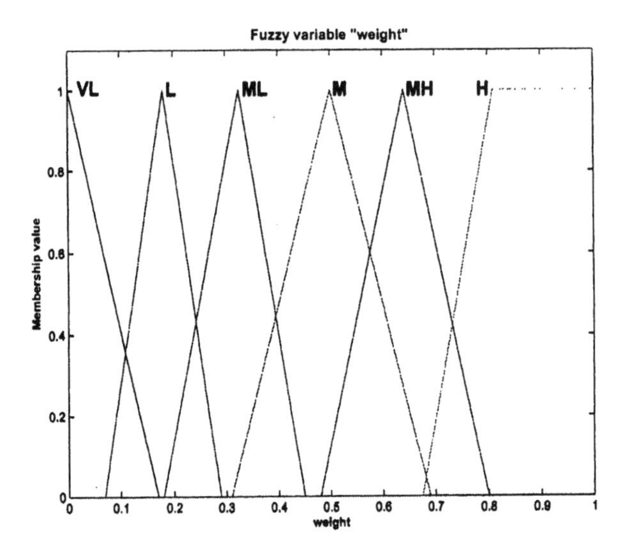

Figure 7.5. Output variable *weight* fuzzy values and corresponding membership functions

mapping is achieved through fuzzy *If...Then* rules, which associate the fuzzy inputs with FIS outputs.

As already mentioned, the implemented demonstrator has two inputs (*time* and *frequency*) and one output (*weight*). The fuzification of the input variables is realized by the use of singletons, and the fuzzy rule base, which comprises 15 rules, provides the desired output[2].

These fuzzy rules are:

- For *frequency*=normal:

```
IF frequency IS normal AND time IS VL THEN weight IS VL
IF frequency IS normal AND time IS L THEN weight IS L
IF frequency IS normal AND time IS ML THEN weight IS ML
IF frequency IS normal AND time IS M THEN weight IS MH
IF frequency IS normal AND time IS H THEN weight IS MH
```

[2]The COA fuzzifier is used

- For *frequency*=med:

```
IF frequency IS med AND time IS VL THEN weight IS L
IF frequency IS med AND time IS L THEN weight IS L
IF frequency IS med AND time IS ML THEN weight IS MH
IF frequency IS med AND time IS M THEN weight IS H
IF frequency IS med AND time IS H THEN weight IS MH
```

- For *frequency*=high:

```
IF frequency IS high AND time IS VL THEN weight IS L
IF frequency IS high AND time IS L THEN weight IS ML
IF frequency IS high AND time IS ML THEN weight IS MH
IF frequency IS high AND time IS M THEN weight IS H
IF frequency IS high AND time IS H THEN weight IS MH
```

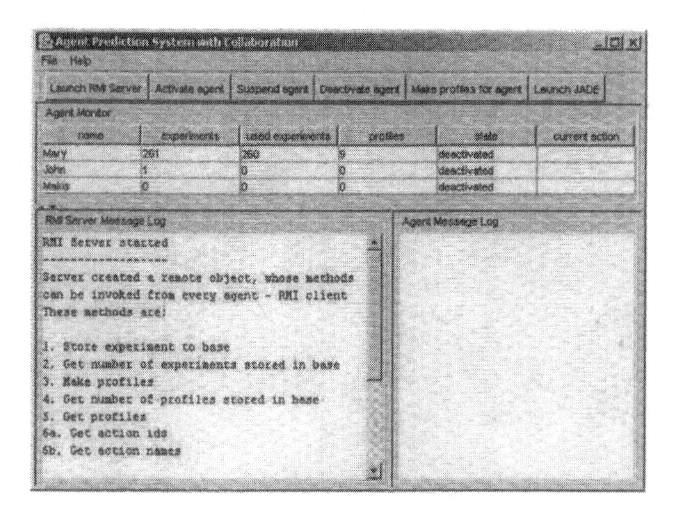

Figure 7.6. The main console of the prototype recommendation engine

The specification of the above rules and the proper formulation of the variables is the only preprocessing task conducted. After data has been processed through the FIS, data mining (the κ-Profile mechanism) is applied.

2.3 Browsing through a Web Site

The developed demonstrator follows the methodology defined in Chapter 5. An RMI server has been employed for realizing all the prediction mechanism tasks, while the browser agents are RMI clients. The application is controlled through an interface console, where the application developer/user can monitor both the server and the clients (Figure 7.6).

When an agent is activated, a web browser is initiated for the user to traverse the web. The user-assigned agent monitors all user actions and identifies web pages it has been "trained"on. In the prototype, the agent has been trained on the http://www.wipeout.gr web site, a commercial music store web site.

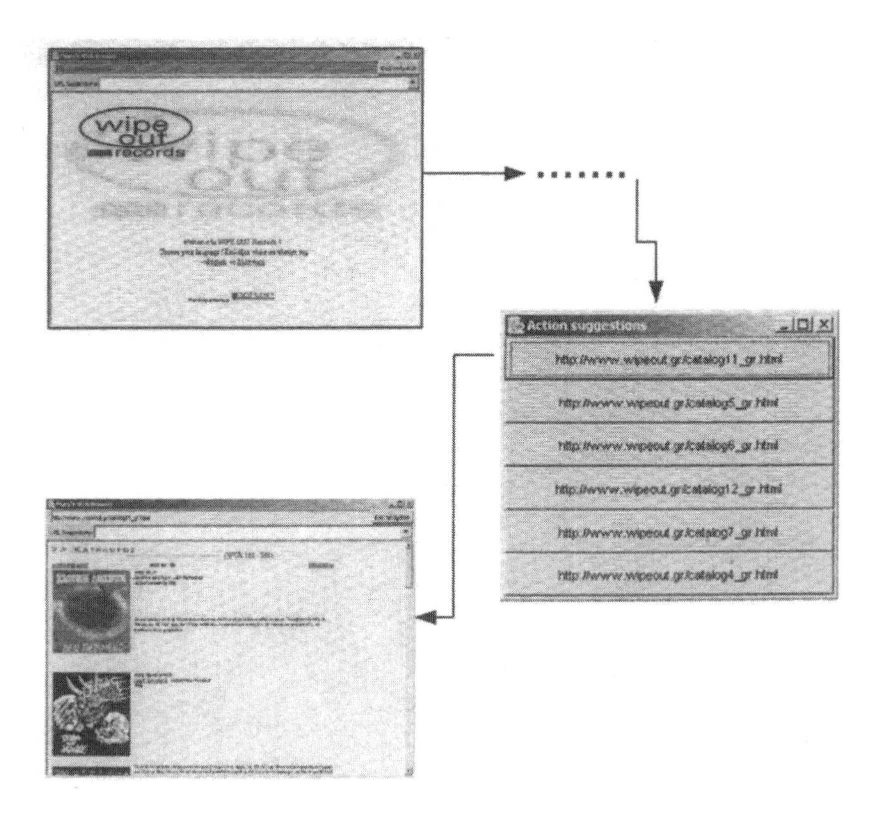

Figure 7.7. The agent recommends a list of possible links of interest, based on its knowledge and the user's prior actions

Figure 7.7 illustrates how the agent, after having navigated through a number of web pages in the "wipeout" site (the recommendation window has been set to $z = 3$), suggests to the user a number of possible links of interest.

In the case the agent cannot make a good suggestion (action recommendation scores are too low), it queries all the other active agents for suggestions, and returns the optimal ones to the user.

3. Experimental Results

The set of action bundle vectors, B, contained 208 elements (b vectors). This set was provided as the training set to the κ-Profile mechanism. At first the *maximin* algorithm identified $K = 8$ clusters, and along with the most representative vectors of each cluster, the $K--Means$ algorithm was applied. The resulting clusters, along with the corresponding set percentage, are illustrated in Table 7.8. As can be easily seen, some of the produced clusters have very low vector membership. This is a result of the layout of the specific web site.

Table 7.8. The resulting vector clusters and their percentage distribution

Cluster	Vector
Cluster 1	56.73%
Cluster 2	4.33%
Cluster 3	~2%
Cluster 4	~3%
Cluster 5	24.5%
Cluster 6	2.4%
Cluster 7	1.44%
Cluster 8	4.33%

Using the resulting clusters, κ-Profile has identified the most representative actions, therefore constructing eight action clusters, which in fact comprise the agent profile set. Based on these profiles, the recommendation engine produced, in real-time, the agent suggestions. Table 7.9 illustrates the actions that comprise the profile of cluster 4, along with their normalized weights within the cluster.

The $WAVP$ metric was applied on the extracted profiles, in order to identify their efficiency. The $WAVP$ metric was calculated to be 78% for the first of the eight profiles extracted, 72.5% for the second profile, while for the rest of the profiles it stayed above a respectable 67%, as illustrated in Figure 7.8.

Table 7.9. The actions that comprise the profile of cluster 4

Action	Vector
p_{67}	0.9998
p_{86}	0.9999
p_{15}	0.8352
p_{82}	0.7827
p_{13}	1.0
p_{11}	0.9788
p_{10}	0.7273
p_9	0.8992
p_{100}	0.8264
p_{77}	0.7892
p_8	0.9999
p_{76}	1.0
p_7	0.9999
p_4	0.8307

As far as the efficiency of the prediction mechanism is concerned, the *PSSR* metric was calculated to be 72.14%, a success rate that was considered satisfactory, since the action space was $n = 142$, while the maximum profile size was $m = 8$ (8 actions maximum). Taking under consideration the fact that the recommendation window was set to $z = 3$ (last three actions of the agent), in almost three out of four cases, the prediction mechanism proposed an action that the agent subsequently followed (a web page that the user chose to visit).

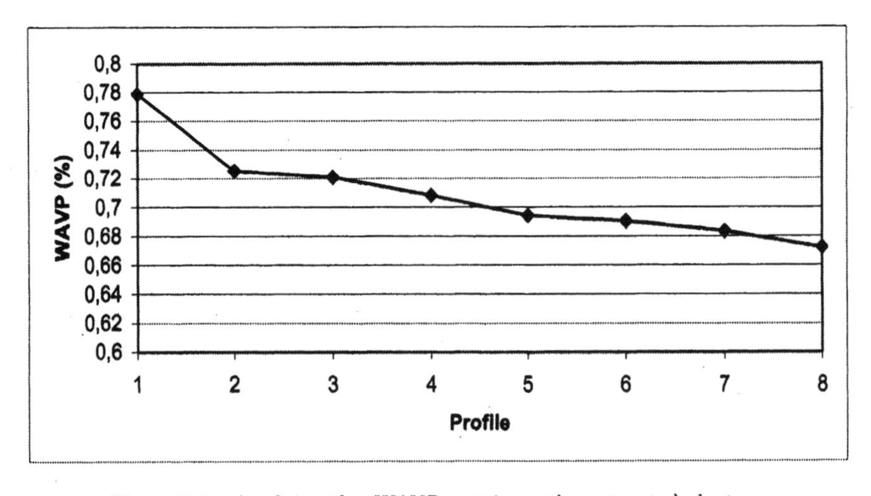

Figure 7.8. Applying the *WAVP* metric on the extracted clusters

4. Conclusions

The main objective of this Chapter was the analysis of all issues involved in the development (through the methodology presented in Chapter 5) of a system that exploits the results of data mining on agent behavior data, in order to predict their posterior actions. A recommendation engine, where agents are web browsers, assisting users in their navigation through a web site, was selected as representative test case, since the direct analogy of an agent action to a user action, could clarify possibly vague points. Through the analysis conducted, we have shown that data mining techniques can, indeed, be exploited for discovering common behavioral patterns between agents, as long as the problem is modeled appropriately and the system infrastructure entails the features presented in Chapter 5. This way data preprocessing is possible and a suitable DM mechanism can be applied, for agent behavior profiles to be extracted. From that point on, agent behavior can be predicted.

Chapter 8

MINING KNOWLEDGE FOR AGENT COMMUNITIES

The third application level of knowledge diffusion differs substantially from the first two, because in this case there are no historical data to perform data mining on and extract knowledge models. Both the multi-agent systems and the DM techniques are evolutionary. Interactions between the members of the system and the influence that some of them exert on others, play a significant role in the overall behavior of the system. We refer to such systems as **agent communities**, because the agents are faced with common problems (or goals), which they attempt to overcome (or reach) through either collaboration or competition. In either case, agent interaction is a pivotal element of the system.

As already described in Chapter 5, key issues for agent communities are the proper modeling of the problem and the efficient representation of the agent knowledge model. As a typical example of this class of MAS, we have chosen to present the simulation of an ecosystem. Agents in an ecosystem, acting as living entities, are born, grow, give offspring, and finally die within the confines of an unknown, possibly hostile environment. Through their actions and their interactions, and by the use of evolutionary DM techniques, the agents can dynamically improve their intelligence, in order to adapt to the ecosystem, and survive.

1. Ecosystem Simulation

Problems related to the analytical modeling of ecosystems are inherently complicated, since the latter are multi-parametrical and inhomogeneous, while the interdependencies between the ecosystems' components are not immediately evident. Nevertheless, these problems are particularly interesting, especially for biologists, software engineers and evolutionary computing researchers, who attempt to model the underlying evolutionary mechanisms from different perspectives due to their different scientific backgrounds. On one hand, biologists usually apply probabilistic techniques, in order to model the structures and validate a pre-specified hypothesis. On the other hand, software engineers and AI researchers use more artificial life-oriented approaches, focusing mainly on the issues of adaptation, self-organization, and learning. In any case, such unilateral methodologies lead to the development of suboptimal solutions.

The flourishing of the AOSE paradigm has provided both biologists and computer scientists with a powerful tool for modeling such complex, nonlinear, evolutionary problems. Using the AOSE approach an ecosystem can be viewed as a network of collaborative, yet autonomous, units that regulate, control and organize all distributed activities involved in the sustainability and evolution of the environment and its elements. Research literature on intelligent agent system architectures has proven that problems that are inherently distributed or require the synergy of a number of distributed elements for their solution can be efficiently implemented as a multi-agent system (MAS) [Ferber, 1999].

Thus, agents are modeled as the individual living organisms of the ecosystem, and are engaged in well-known characteristic vital activities – they live, move, eat, communicate, learn, reproduce, and finally die. The other non-agent parameters of the ecosystem are specified within the context of the agent society and, in turn, have a profound effect on the community.

Numerous research initiatives exist, where agents and agent communities have been developed for simulating ecosystems and their various aspects. In [Haefner and Crist, 1994; Crist and Haefner, 1994; Werner and Dyer, 1994], researchers have built individual-based models that are focused on the environmental aspects of the systems, while some models consider both ecological and evolutionary dynamics, in communities composed of many interacting species [Caswell, 1989; DeAngelis and Gross, 1992; Durrett, 1999; May, 1973; Pecala, 1986]. Bousquet et. al. [Bousquet et al., 1994], on the other hand, have focused on the societal aspects of a model, where a population of agents representing humans exploits ecological resources. Advancing to more artificial life-oriented approaches, the main focus is given on the societal impact of the evolution and on the learning processes employed. Such models vary from relatively simple ones [Ackley and Littman, 1990; Yaeger, 1994], to more elaborate Complex Adaptive Systems (CAS) [Holland, 1995], like Swarm [Langton, 1994], Sugarscape [Epstein and Axtell, 1996], Tierra [Ray, 1992], and Echo [Hraber et al., 1997].

Special attention should be drawn, however, to the work of Krebs and Bossel [Krebs and Bossel, 1997], who have developed the model of an artificial animal ('animat') [Wilson, 1990; Dean, 1998] residing in an unknown environment. Within the context of their work, the authors have successfully integrated the environmental aspects of ecosystems with an efficient mechanism for learning, thus providing an efficient self-organizing animat having to cope with complex environments.

Further elaborating on this work we have developed *Biotope*, an agent community for simulating an ecosystem, where agents learn how to interact with an unknown, possibly hostile environment. The main objectives of Biotope were: a) to develop a fully-functional, highly-configurable tool for designing ecosystems, for simulating them, and monitoring evolution through certain environmental indicators, and b) to study the improvement of the agent community learning curve with respect to environmental and agent characteristics, by modifying the parameters that are related to agent communication, knowledge exchange, and self-organization.

The key improvement points of Biotope, with respect to the work presented by Krebs and Bossel are:

1. Communication issues have been introduced. Ecologists have, in various ways, acknowledged the fact that in most populations individuals interact more with their neighbors and immediate surroundings, than they do with remote individuals and structures [Westerberg and Wennergren, 2003]. In order to take this assumption into account, we have replaced the notion of the animat with that of an agent, that can communicate with its neighbors and exchange information on the environment.

2. The learning mechanism employed has been improved, in order to better reflect the needs of the agents with respect to their environment. The classifier system [Wilson and Goldberg, 1989; Booker et al., 1989; Wilson, 1987] implemented has been refined, while the genetic algorithm genotypes [Goldberg, 1989; Holland, 1987; Holland, 1975] have been revised. In addition, the parameters of the learning mechanism (genetic algorithm application, rule exchange) are reconfigurable.

3. The Dispersal Distance Evolution theory [Murrel et al., 2002; Rousset and Gandon, 2002] has been introduced to the system. Agent reproduction, which is now introduced, follows the primitives of this theory, in order to ensure a more concise, near-real ecosystem model.

4. The environment has been enriched with more elements to offer a greater variety and we have realized a number of environmental assessment indicators based on the orientation theory [Bossel, 1977].

5. The vision field of an agent has been expanded from 3×3 to $n \times n$, while the concept of aging has been introduced.

The rest of this chapter is organized as follows: Section 2 describes the implemented ecosystem infrastructure, specifying the characteristics of its structural elements. The environment and the agent model are analyzed, whereas a detailed overview of the learning mechanism is provided. Section 3 illustrates the basic functional operations of Biotope, and presents a typical scenario of use. Finally, section 4 goes over three series of experiments conducted with Biotope: a) increasing agent intelligence by agent interaction, b) improving the agent learning mechanism in unreliable environments, and c) assessing overall environment quality.

2. An Overview of Biotope

2.1 The Biotope Environment

The environment is defined as a two-dimensional ($x \times y$) grid with no movement permitted beyond its borders. Each one of the grid cells represents a site of the environment and can be referenced through its coordinates. The current implementation of Biotope employs a two-dimensional integer array for mapping environment points.

A cell in the Biotope environment may contain food, a trap, an obstacle, an agent, or it may be vacant. Table 8.1 shows the numerical values associated with each cell type, while Figure 8.1 illustrates an example of a 15×15 environment. When generating a new ecosystem, the user defines the population of each one of the environment objects and Biotope scatters them on the grid following a normal distribution. Agents are randomly dispatched, in vacant cells in the grid.

Table 8.1. Mapping the contents of Biotope

Cell value	Cell content
0	Vacant space
1	Food
2	Trap
3	Obstacle
4	Agent

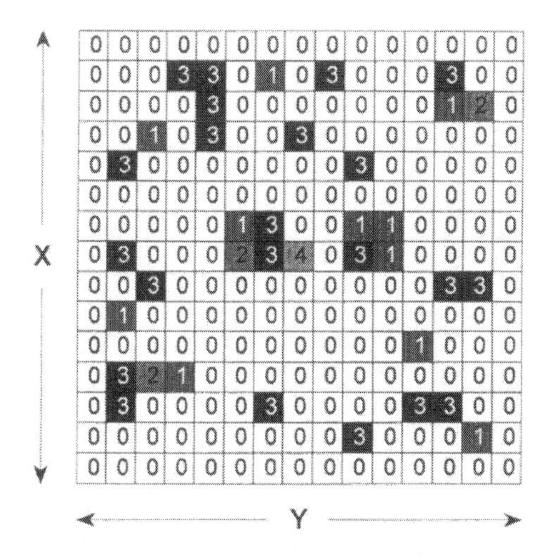

Figure 8.1. An overview of the Biotope environment

2.2 The Biotope Agents

Agents in Biotope represent the living organisms of the ecosystem that have a dual goal: to survive and multiply. In order to achieve this, agents have to learn how to move and behave in the, initially unknown, environment and to increase their energy, in order to produce more offspring throughout their lifetime. Biotope agents augment their intelligence by employing a learning mechanism that involves exchanging knowledge with their neighbors on the contents of their surroundings.

2.2.1 Agent sight

Instead of using the von Neumann or the Moore neighborhood, Biotope agents have the ability to "see" an $m \times m$ area of the environment (extended Moore neighborhood). The notion of uncertainty has been introduced into Biotope, with the addition of an error probability p to agent vision. The user can, therefore, specify the environmental reliability (i.e. the probability that what the agent sees is what the agent gets), by altering p. For each agent, a vision vector of $m \times m - 1$ values is constructed by taking, in a row-major fashion, the contents of the cells around the agent's position. Figure 8.2 illustrates a 5×5 vision field and its corresponding vision vector.

2.2.2 Agent movement

Agents can move towards any cell within their vision field, as long as this cell is not occupied by another agent or any obstacle. The classifiers

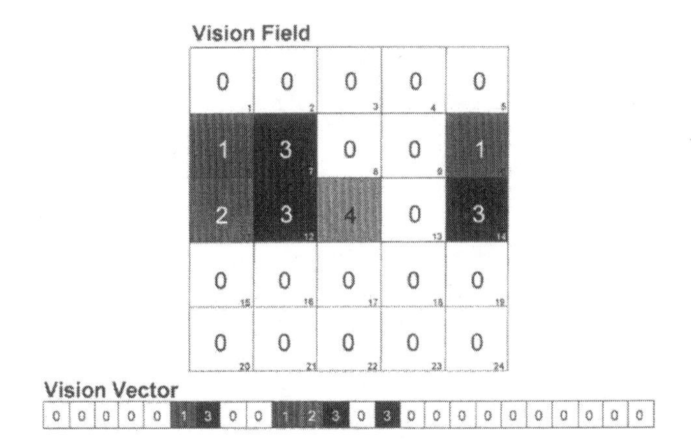

Figure 8.2. Agent vision field and the corresponding vision vector

(movement rules) of the agent decide on the destination cell. The vision vector of the agent is compared to the set of classifiers and, if a match occurs, the agent moves toward the proposed cell (Figure 8.3).

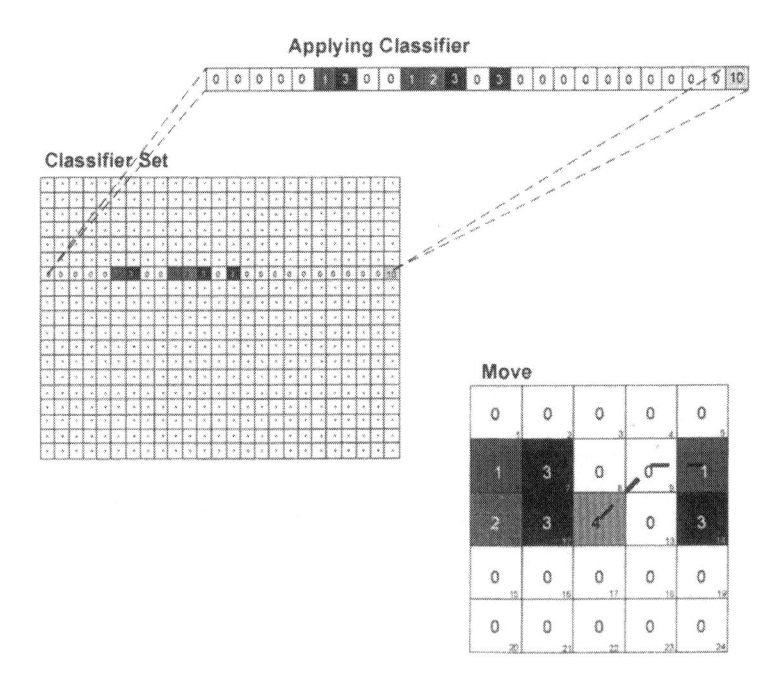

Figure 8.3. Deciding on the next move, based on the classifier set

The route that the agent will follow towards its destination is cost dependent. All possible paths are considered and the shortest one is selected (Figure 8.4), based on the maze algorithm [Lee, 1961]. Traps, obstacles and agents are the "walls" of the maze that the agent has to overcome.

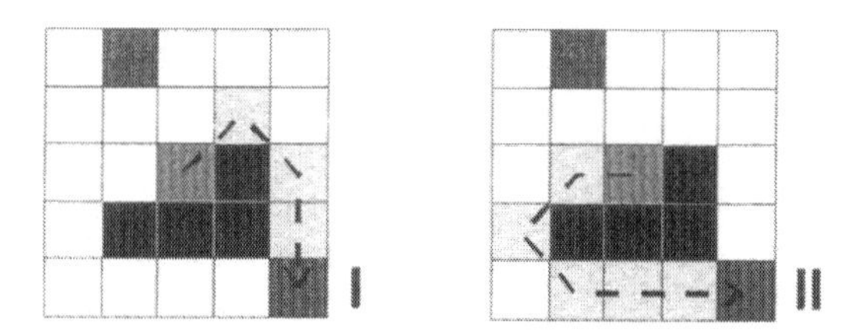

Figure 8.4. Two possible paths towards the destination cell: The first route is selected, since the energy cost is less

2.2.3 Agent reproduction

An agent can reproduce, iff its energy balance is equal or greater than z energy units (eu). Only one offspring is born each time and it inherits part of the parent's perception of the environment. Following the dispersal theory primitives, the birth location and the initial energy balance of the newly born, the decrease of energy of the parent, and the percentage of the classifier set transferred are determined by the following probability density function:

$$Exp(x) \sim \left(\frac{1}{d_s}\right) \times \exp\left(\frac{-x}{d_s}\right) \tag{8.1}$$

where x is the dispersal distance at birth and d_s is a constant defined as the quarter of the grid's diagonal. It should also be denoted that the dispersal orientation is random.

2.2.4 Agent communication – Knowledge exchange

Agent communication constitutes an important factor which is thoroughly exploited in this work. The messages exchanged aim to ensure agent collaboration and knowledge diffusion from the "wiser" to the more "naive" agents. In order to establish communication, two agents must fall within each other's vision field. FIPA-compliant [The FIPA Foundations, 2000] messages are issued according to the protocol illustrated in Figure 8.5. A message may contain an agent's entire classifier set or a percentage of it, specified during the modeling of the ecosystem. The stronger rules transferred replace the weaker rules of each agent's knowledge base. Once communication between two agents is completed, it cannot be re-established until a pre-specified number of epochs has elapsed, in order to avoid stateless messages between agents residing in the same neighborhood.

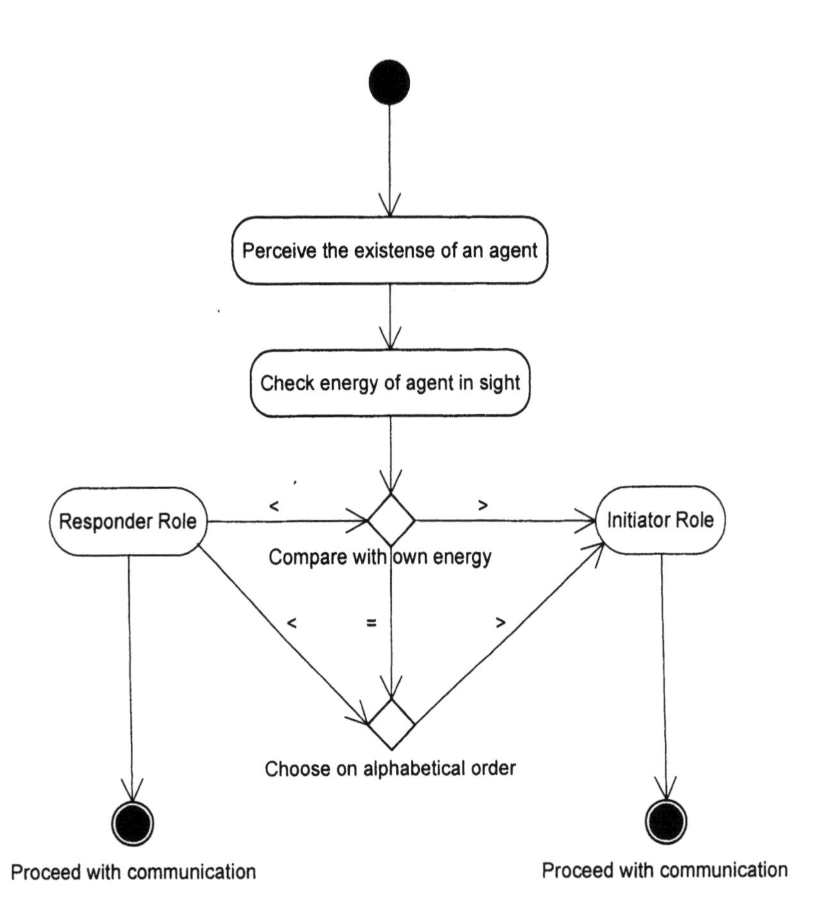

Figure 8.5. Establishing communication between neighboring agents

2.3 Knowledge Extraction and Improvement

One of the basic features of an ecosystem developed with Biotope is the ability of its members to augment their intelligence. We have implemented a learning mechanism that comprises three parts: i) the set of classifiers (rules), ii) a classifier evaluation mechanism, and iii) a genetic algorithm implementation.

2.3.1 Classifiers

The classifiers are the agents' driving force within the environment. Each classifier consists of two parts: the detectors and the effectors and can be represented as in Eq. 8.2:

$$If \ < detector > \ then \ < effector > \tag{8.2}$$

This type of classifier can be directly mapped to the structure of an agent message:

$$< Classifier\ rule >::=< condition >:< movement > \tag{8.3}$$

where the $< condition >$ clause is a stream of 0, 1 and wildcards (#), while the $< movement >$ clause is a stream of 0 and 1.

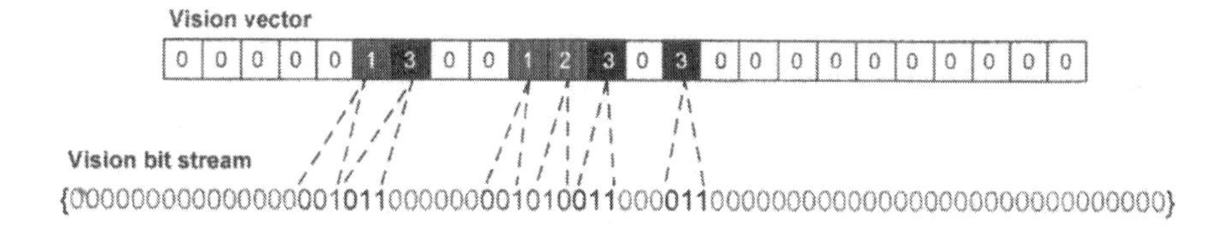

Figure 8.6. Transforming the vision vector into a bit stream

Every time an agent moves to a new position within its vision field, a new vision vector is generated. This vector is transformed into a bit stream (Figure 8.6), which facilitates comparison to the effectors of the set of classifiers.

When a detector stored in the agent's knowledge base matches the vision bit stream, then the detector is activated and the agent moves towards the direction it imposes. The initial classifiers are constructed with respect to the agent's wandering path. If no classifier detector matches the current agent position, then a new classifier is constructed, with the vision bit stream as a detector, and a random effector. Table 8.2 illustrates how the perceived environment (through agent messages) is transformed into agent moves.

Since it is quite probable to have more than one classifier matching the incoming vision bit stream (see Table 8.2), we have employed an enhanced version of the classifier evaluation mechanism.

2.3.2 Classifier Evaluation mechanism

A slightly modified version of the bucket brigade algorithm is used for evaluating classifiers. Each time there are multiple matches, a bidding

Table 8.2. Perceiving the environment and taking action

Agent Message	Classifier Rule	Agent Movement
000000101020020000001	No match - Generate new rule	Random location (e.g. Cell 8)
010000002030030000001	0#0000#0#0300300000001:24	Cell 24
000000001030030000001	0#0000#0#0300300000001:24	Cell 24

process between the candidate classifiers is initiated. Bidding is based on the strength of each classifier rule, a quantity that represents the success rate of a classifier with respect to its prior decisions (i.e., when a classifier rule sends an agent to a food cell, then its strength is increased and vice versa). The bid that a classifier C participating in an auction is going to place at time t, is given by:

$$bid(C,t) = C_{bid} \times strength(C,t) \tag{8.4}$$

where C_{bid} is a constant value that determines the strength percentage each classifier is willing to place in the bid. The classifier that places the higher bid is pronounced winner of the auction and it is activated. The strength for C is now (at time $t+1$):

$$\begin{aligned} strength(C,t+1) &= strength(C,t) + energy(C,t+1) - \\ &- energy(C,t) - bid(C,t) - tax(C,t) \end{aligned} \tag{8.5}$$

where $energy(C,t+1)$ is the current energy of the agent (after having followed the route to the indicated cell and having consumed the food there, or having fallen into the trap), $energy(C,t)$ is the energy of the agent on time t, and $tax(C,t)$ a value for participating in the auction (all participating classifiers pay the tax). Finally, C rewards the classifier C' that was activated before (at time $t-1$) and led to the activation of C. The new strength value of C' is now:

$$strength(C',t+1) = strength(C',t) + bid(C,t) - tax(C,t) \tag{8.6}$$

2.3.3 Genetic Algorithm

In order to increase the variety of classifiers, we employ a genetic algorithm mechanism that generates new classifiers. These classifiers replace the weaker (in terms of strength) of the classifiers in the knowledge base of the agents, providing therefore a notion of "rejuvenation" of their mentality. The frequency of this process, which is illustrated in Figure 8.7, is defined through the Biotope user interface.

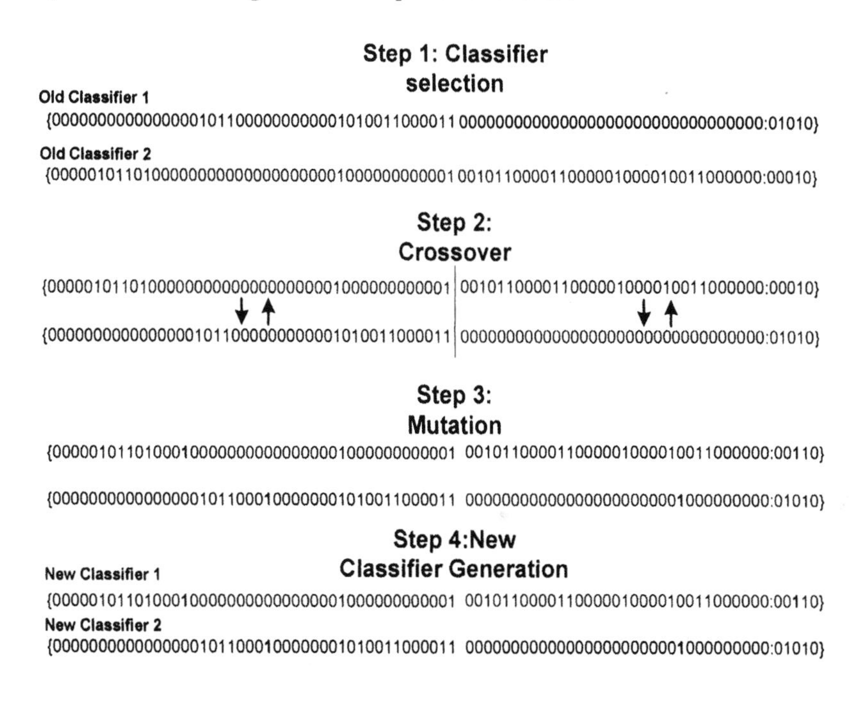

Figure 8.7. Creating new Classifiers

2.4 The Assessment Indicators

One of the primary objectives of the developed system is to provide an analytical description tool for monitoring both the environmental quality and the improvement in the behavior of the agents in the ecosystem. Biotope includes a series of environmental and agent assessment indicators, for the user to better comprehend how changes in the ecosystem reflect to its members, and how changes in the behavior of the agents affect their environment.

2.4.1 Environmental indicators

Resource availability. Resource availability a is defined as the ratio of the total food f available in the environment to the total harvesting distance d_f:

$$a = \frac{\sum f}{\sum df} \tag{8.7}$$

The denominator of Eq. 8.7 is calculated as the sum of steps needed in order to collect all food cells, starting from a random position and moving always to the nearest food cell (Traveling Salesman problem – [Lawler et al., 1992]).

Environmental variety. The Environmental variety v of an ecosystem is defined as the ratio of the distinct vision vectors vv that may exist to the occupied cells of the grid oc (cells that contain either food, trap or obstacle):

$$v = \frac{\sum vv}{\sum oc} \tag{8.8}$$

Environmental reliability. Environmental reliability is defined as:

$$r = 1 - p \tag{8.9}$$

where p is the vision error probability, that is, the probability that an agent will misinterpret the vision vector.

2.4.2 Agent performance indicators

Energy. Energy En is they key feature of an agent. At birth, agents have an initial energy level, which goes up or down, depending on their actions (eating food increases energy, falling in traps and colliding with obstacles reduces it, moving burns it, etc.). When an agent runs out of energy, it dies. Table 8.3 illustrates the correlation between agent actions and the energy variation rate.

Table 8.3. Agent actions and energy variation rate

Agent Action	Energy Variation Rate
Agent movement	$-1\ eu/step$
Food consumption	$10\ eu/foodcell$
Trap collision	$-10\ eu/trap$
Agent reproduction	$x\ eu/reproduction$ (determined by the dispersal distance theory)

It should be denoted that the energy loss rate increases, when the organism "ages", i.e., it exceeds a pre-specified number of epochs.

Effectiveness. Effectiveness e is defined as:

$$e = 1 + \frac{eur - elr}{ear} \tag{8.10}$$

where eur is the energy uptake rate, elr the energy loss rate and ear the energy availability rate, as defined in [Krebs and Bossel, 1997].

Aging. An integer ag_i defines the age of agent i, measured in ecosystem epochs. As already mentioned, aging increases the energy loss rate.

Food consumption rate. The food consumption rate indicator fcr is defined as:

$$fcr = \frac{\sum f_{encounter}}{s} \tag{8.11}$$

where the numerator represents the number of food cells the agent hits, while the denominator represents the moves that the agent has performed.

Trap collision rate. Trap collision rate is defined as:

$$tcr = \frac{\sum t_{collision}}{s} \tag{8.12}$$

where $\sum t_{collision}$ is the number of traps the agent has collided with.

Unknown situation rate. The unknown situation rate usr is defined as:

$$usr = \frac{\sum u_{situation}}{s} \tag{8.13}$$

where $\sum u_{situation}$ is the number of total unknown situations (no classifier matches the vision vector) that the agent has encountered.

Reproduction rate. Finally, the agent reproduction rate rr is defined as:

$$rr = \frac{\sum r_{reproduction}}{s} \tag{8.14}$$

where $\sum r_{reproduction}$ is the number of offspring the agent has given.

3. The Implemented Prototype

Biotope provides a multi-functional user interface, to facilitate researchers in their experiments. The system has been implemented in Java v.1.4.2 and all the agents are developed over the JADE v.3.0, which conforms to the FIPA specifications.

Table 8.4 illustrates the functional components of the main application window, while Table 8.5 shows the functionalities of the application menu bar:

Table 8.4. The application components of Biotope and their functionalities

Application Component	Functionalities
Application Toolbar	- Run/Pause/Stop simulation - View grid: Observe the environment in real-time - Focus off: Stop monitoring a selected agent
Ecosystem monitoring panel	- Monitor all agents of the platform - Monitor information on the indicators of each agent - Focus on agent
Message monitoring panel	- View all the control messages of Biotope - View births and deaths - View knowledge exchange
Indicator panel	- View the environmental indicators - View statistical info

Table 8.5. The application menu bar items

Submenus	Submenu items
File	- Open Grid: Opens a stored experiment configuration - Save Grid: Saves an experiment configuration - Exit: Terminates Biotope
Settings	- Environment: Opens the environment configuration dialog - Agents: Opens the agent configuration dialog
Actions	- Grid - Focus off
View	- Launch JADE: Opens the JADE console - Run/Pause/Stop
Help	- About

3.1 Creating a New Simulation Scenario

This section illustrates how to configure and run a new simulation. When Biotope is launched, the main application window appears (see Figure 8.10).

From the "Settings" menu, the user can configure the environment parameters (grid dimensions, food, trap, and obstacle values, agent vision error percentage, food refresh rate, initial population – Figure 8.8), as well as the agent parameters (initial energy, aging step, knowledge base size, genetic algorithm step, communication step, exchanged rules percentage – Figure 8.9).

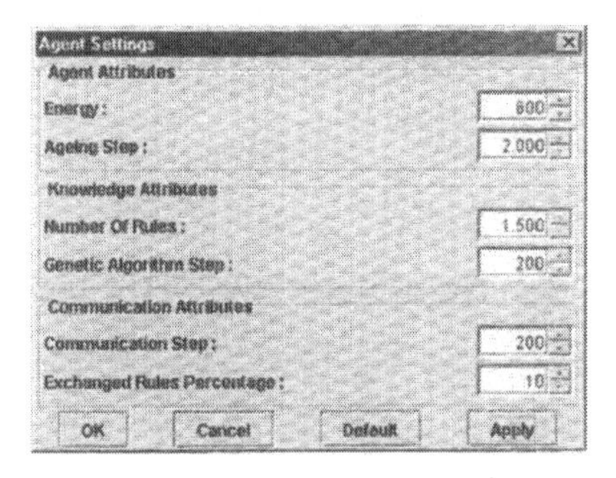

Figure 8.8. Configuring environmental parameters

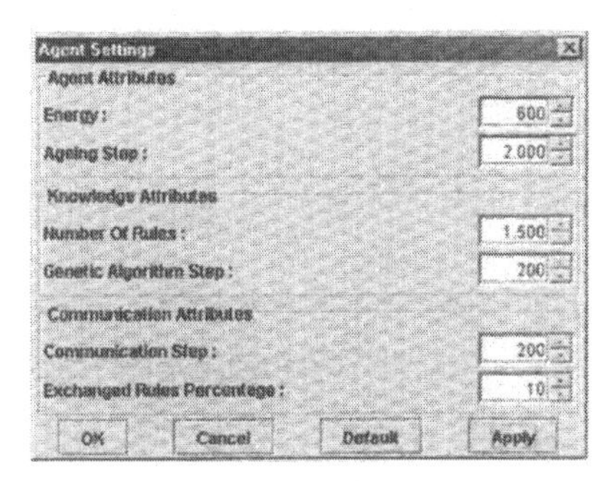

Figure 8.9. Configuring agent parameters

From the "View" menu, the "Grid" option can be selected, in order to visualize the experiment and the experiment is then initiated (the "Run" item from the "Actions" menu is selected). Figure 8.10 illustrates a snapshot of Biotope "in action".

All agent indicators are presented on the main application window, to enable the user to monitor the simulation at real time, while all the

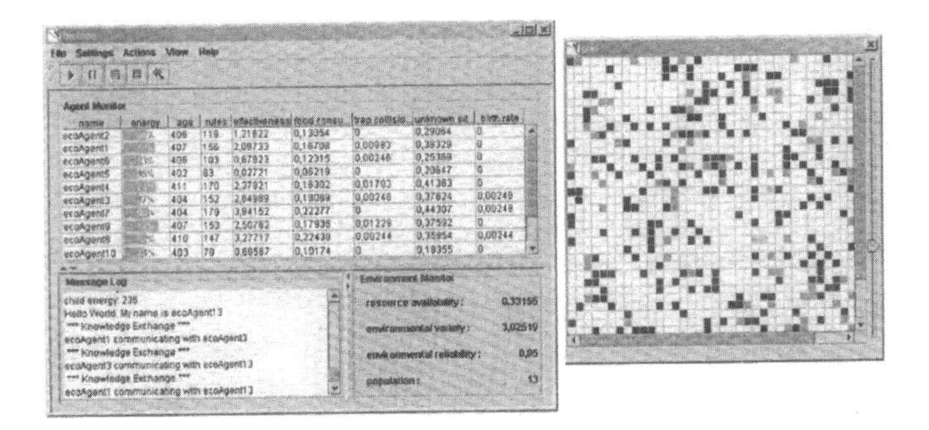

Figure 8.10. Biotope "in action"

variables are recorded in appropriate vectors, for posterior processing. The user can focus on one of the agents, by clicking on it, or he/she can start the JADE Sniffer, to observe the communication between agents.

When the user decides that the system has reached a certain equilibrium (by monitoring the changes in the indicators), he/she can stop the simulation process, and store the experiment for future reference.

4. Experimental Results

We report here the results obtained for three different series of experiments performed with Biotope, in order to evaluate its capabilities and demonstrate its advantages. The first series of experiments ($E_{A-1} - E_{A-8}$) illustrates how agent communication parameters can be tuned, in order to improve the performance of multi-agent communities residing in relatively reliable ecosystems. In the second series of experiments ($E_{B-1} - E_{B-6}$), the environment reliability is significantly reduced. It turns out that in unreliable environments, agent communication proves insufficient to lead the system to equilibrium, and the necessity of genetic algorithms becomes evident. Genetic algorithm parameters are specified, in order to show the added-value of the optimization mechanism. Finally, using near-optimal values for the parameters of agent communication and genetic algorithm, a third series of experiments was conducted ($E_{C-1} - E_{C-10}$), simulating various environments. Interesting conclusions were drawn and are further discussed.

The environment and agent parameters that have remained constant through all the experiments for comparison reasons are listed in Table 8.6.

All the indicators were recorded, with special attention given to agent population, effectiveness e, food consumption rate fcr, trap collision rate tcr, unknown situation rate usr, and reproduction rate rr.

Table 8.6. Fixed parameter values for all the experiments

	Parameter	**Value**
Environmental parameters	Grid Dimensions	x : 30 y : 30
Agent parameters	Vision Field	5×5
	Initial agent energy	500 *eu*
	Aging	3000 *epochs*

4.1 Exploiting the Potential of Agent Communication

In order to study the impact of agent communication in unknown environments, we have performed eight experiments with different communication parameters. The initial population of the ecosystem was set to 10 agents. Food refresh rate was set to 1/40 *epochs*. The values for the Environmental variety (v) and Resource availability (a) indicators were calculated to $v = 0.319081$ and $a = 2.10345$, respectively. Table 8.7 summarizes the parameters for each experiment.

Table 8.7. Experiments on agent communication

	Parameters			
Experiment	Communication Step	Exchanged Rules	Knowledge Base size	Vision Error
E_{A-1}	1000	1%	3000	5%
E_{A-2}	500	10%	3000	5%
E_{A-3}	200	10%	3000	5%
E_{A-4}	50	10%	3000	5%
E_{A-5}	200	10%	1000	5%
E_{A-6}	200	10%	200	5%
E_{A-7}	200	10%	3000	20%
E_{A-8}	100	25%	3000	20%

Our efforts focused on three different aspects of agent communication, with respect to learning: a) to determine, for a given system, the optimal communication rate for agents, b) to study the effect of the agents' knowledge base size on the agent indicators, and c) to study the communication efficiency of the community when environmental reliability decreases. In the following analysis, representative results for the most interesting cases are presented.

4.1.1 Specifying the optimal communication rate

While increasing agent communication rate, ranging from 1 *exchange/ 1000 epochs* (E_{A-1}) to 1 *exchange/50 epochs* (E_{A-4}), we observe (Table 8.8) that we have an improvement in the average values of the indicators, advancing from E_{A-1} to E_{A-3}. E_{A-4}, however, does not follow the same trend. The graphs in Figure 8.11 compare the performance of the agent community in experiments E_{A-3} and E_{A-4}. Exponential trend lines have been drawn, to delineate differences. Data for experiments E_{A-1} and E_{A-2} have been omitted to allow for better visualization.

Table 8.8. Average indicator values for experiments E_{A-1} to E_{A-4}

Average Indicator Values						
Experiment	Population	e	fcr	$tcr \times 10^{-3}$	ucr	$rr \times 10^{-3}$
E_{A-1}	8.5	1.459	0.135	4.953	0.346	0.309
E_{A-2}	10.8	1.512	0.134	6.326	0.431	0.270
E_{A-3}	13.1	1.641	0.132	5.107	0.418	0.288
E_{A-4}	11.2	1.494	0.126	5.365	0.441	0.262

In all cases the indicator trend lines of E_{A-3} overcome those of E_{A-4}. Given the fact that the genetic algorithm mechanism has been practically disabled (employed once every 5000 epochs), increasing the communication rate to 1 *exchange/50 epochs* seems to introduce redundancy into the system. The same agents (neighbors) communicate with each other in very short intervals, not having any new classifiers to exchange. So, in the environment selected, setting the communication rate to 1/200 *epochs* seems to provide the optimal solution. In a different environment, tuning would be necessary to determine an optimal communication rate, a process easily implemented in Biotope.

4.1.2 Agent efficiency with respect to their knowledge base size

Retaining the basic parameter values specified in E_{A-3}, which produced an optimal evolution scenario for our community, we, then, modified the size of the agents' knowledge base. The results of experiments E_{A-5} and E_{A-6} were particularly interesting, since they led to the obser-

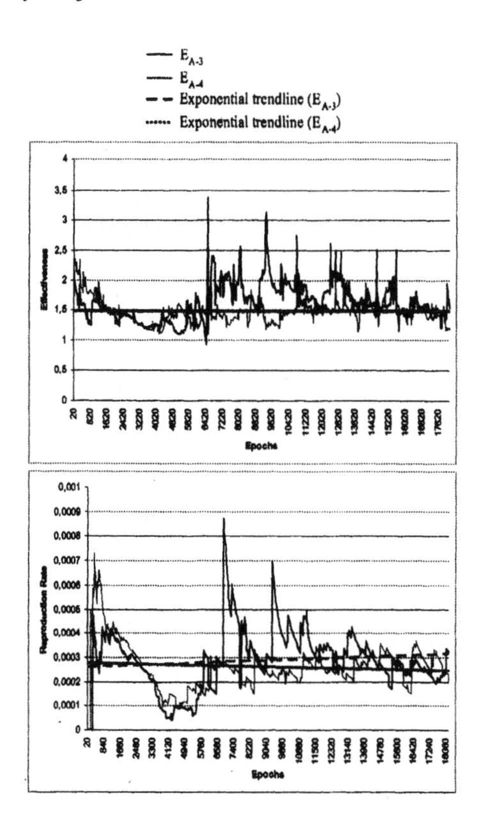

Figure 8.11. Comparing E_{A-3} and E_{A-4} through their agent indicators

vation that the size of the knowledge base can indeed result in different evolution patterns. Table 8.9 lists the average values of the agent indicators for the two experiments, while Figure 8.12 shows the performance of the agent community in the two cases.

In all the graphs the indicator trend lines of E_{A-5} overcome those of E_{A-6}. The difference in the knowledge base size (1000 rules for E_{A-5} compared to 200 rules for E_{A-6}) had a significant effect on the learning process of the community, reinforcing the, nevertheless intuitive, conclusion that more rules mean better perception of the environment.

4.1.3 Exploiting agent communication in unreliable environments

Finally, in E_{A-7} and E_{A-8} we decreased the reliability of the environment, while modifying the percentage of exchanged rules and the communication rate. Table 8.10 illustrates the average indicator values

Table 8.9. Average indicator values for experiments E_{A-5} and E_{A-6}

Average Indicator Values						
Experiment	Population	e	fcr	$tcr \times 10^{-3}$	ucr	$rr \times 10^{-3}$
E_{A-5}	9.2	1.634	0.139	6.280	0.399	0.298
E_{A-6}	9.6	1.392	0.127	6.875	0.406	0.199

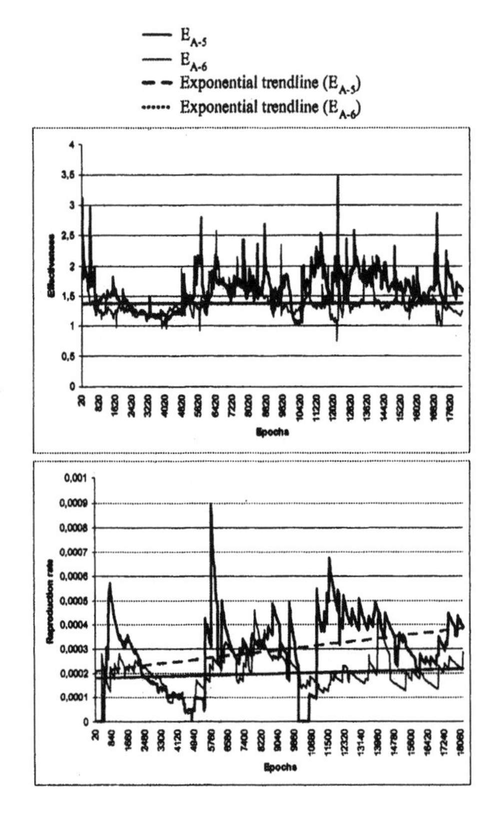

Figure 8.12. Comparing E_{A-5} and E_{A-6}

for these experiments, while Figure 8.13 gives the variance of two agent indicators in both cases.

Although the modification of agent communication parameters has led to an increase in the overall performance of the system, E_{A-7} terminates unexpectedly (all agents die), while E_{A-8} does not reach some equilibrium. This behavior shows that simple rule exchange is not sufficient when the environment is unreliable. In these cases, the genetic algorithm mechanism has to be employed.

Table 8.10. Average indicator values for experiments E_{A-7} and E_{A-8}

Average Indicator Values						
Experiment	Population	e	fcr	$tcr \times 10^{-3}$	ucr	$rr \times 10^{-3}$
E_{A-7}	6.9	1.318	0.128	14.773	0.472	0.194
E_{A-8}	7.6	1.417	0.127	12.020	0.431	0.251

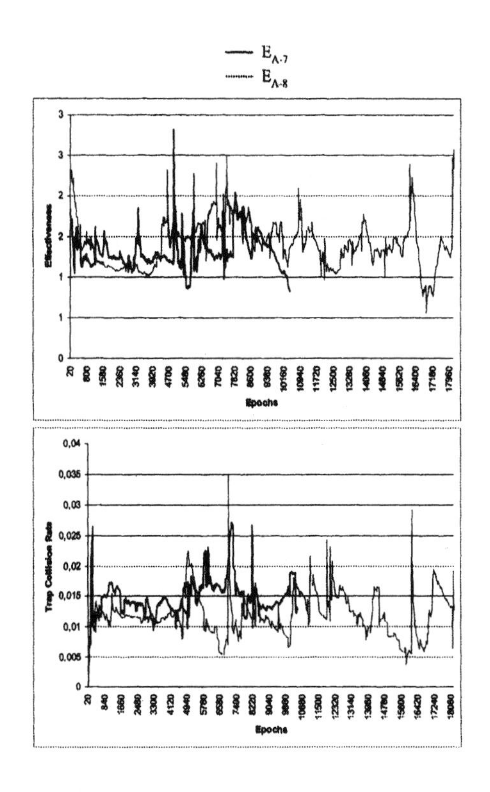

Figure 8.13. Comparing E_{A-7} and E_{A-8}

4.2 Genetic Algorithms for Agents in Unreliable Environments

In experiments E_{B-1} to E_{B-6} we wanted to study the use of genetic algorithms (GA) in unreliable environments. Agent communication parameters were set up so as not to affect the learning process. The initial population of the ecosystem was set to five agents. Table 8.11 shows the parameters of each experiment. It should also be noted that the *Knowledge Base size* was set at 1500 rules and the *Communication step* at 1000 epochs in all experiments.

Table 8.11. Experiments on Genetic Algorithm application

Experiment	Parameters			
	GA Step	Vision Error	Environmental Variety	Resource Availability
E_{B-1}	1000	30%	3.17209	0.34462
E_{B-2}	250	30%	3.17209	0.34462
E_{B-3}	50	30%	3.17209	0.34462
E_{B-4}	1000	50%	3.17209	0.34462
E_{B-5}	250	50%	3.17209	0.34462
E_{B-6}	50	50%	3.17209	0.34462

The experiments were grouped into two categories. Group 1 (E_{B-1} to E_{B-3}) monitored population growth in the same environment, but with varying GA application rate. Group 2 (E_{B-4} to E_{B-6}) compared agent community performance in two different unreliable environments (0.7 and 0.5 respectively), with respect to the same GA application rate (1 application/1000 – 250 – 50 steps).

Figure 8.14 illustrates the dramatic increase in the population of the ecosystem, as the GA application rate increases, while Table 8.12 summarizes the average values of the agent indicators for all the experiments.

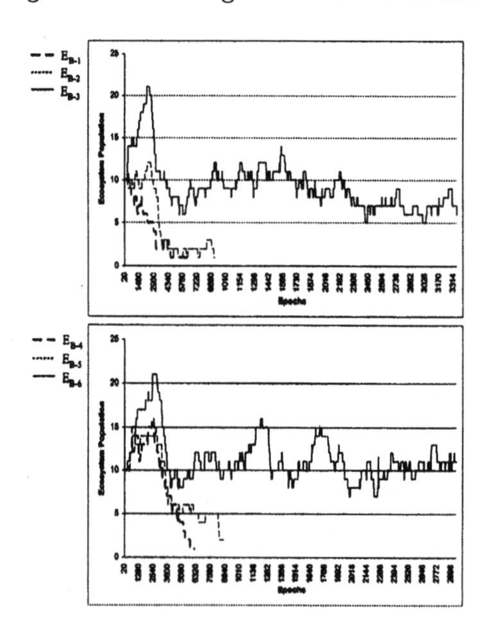

Figure 8.14. Population growth with respect to varying GA application rate

Table 8.12. Average indicator values for experiments E_{B-1} to E_{B-6}

Average Indicator Values						
Experiment	Population	e	fcr	$tcr \times 10^{-3}$	ucr	$rr \times 10^{-3}$
E_{B-1}	9.0	1.247	0.115	9.935	0.460	0.16
E_{B-2}	8.2	1.403	0.119	10.079	0.474	0.18
E_{B-3}	11.4	1.500	0.113	9.673	0.559	0.22
E_{B-4}	4.1	1.207	0.109	14.111	0.483	0.15
E_{B-5}	5.1	1.508	0.106	18.216	0.609	0.25
E_{B-6}	9.3	1.542	0.113	11.672	0.615	0.27

Figure 8.15 compares experiments E_{B-1} & E_{B-4} to E_{B-3} & E_{B-6}, in order to show the convergence of the behaviors in the two populations, when the GA application rate increases. When agents generate new classifiers more often (every 50 epochs), their behavior is improving.

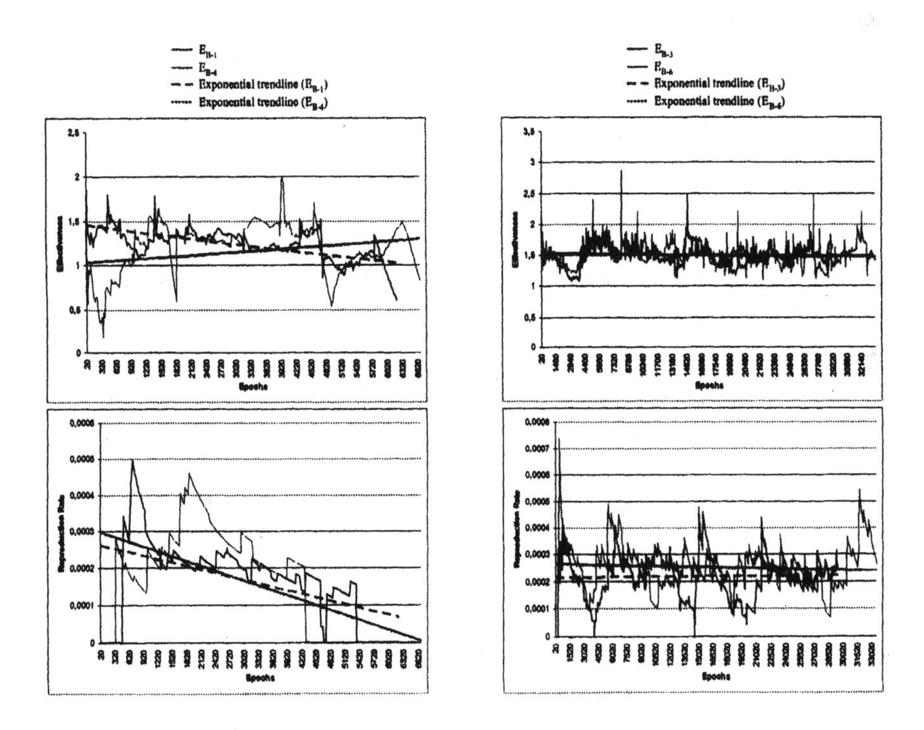

Figure 8.15. Convergence in the behaviors of agent communities when the GA application rate increases

4.3 Simulating Various Environments

Finally, we have conducted a series of experiments for different ecosystems. Having taken the observations on agent communication and learning into account, we have chosen near-optimal values for the parameters of these mechanisms. An initial population of five agents was set, with GA application and agent communication step set at 200 epochs, while the main focus in this study was given to the changes of the indicators with respect to varying resource availability, environmental variety, and environmental reliability. Interesting observations were extracted and are discussed. Table 8.13 presents the ecosystem parameters for each one of the experiments.

Table 8.13. Experiments on various environments

	Parameters			
Experiment	Resource Availability	Environmental Variety	Environmental Reliability	Food Refresh Rate
E_{C-1}	0.35450 - HIGH	3.58974 - HIGH	0.99 - HIGH	35
E_{C-2}	0.33450 - HIGH	3.61026 - HIGH	0.60 - LOW	35
E_{C-3}	0.34050 - HIGH	1.53521 - LOW	0.99 - HIGH	35
E_{C-4}	0.27089 - LOW	3.73158 - HIGH	0.99 - HIGH	35
E_{C-5}	0.27185 - LOW	1.53521 - LOW	0.99 - HIGH	35
E_{C-6}	0.27198 - LOW	3.85405 - HIGH	0.60 - LOW	35
E_{C-7}	0.33909 - HIGH	1.90323 - LOW	0.60 - LOW	35
E_{C-8}	0.26781 - LOW	1.53521 - LOW	0.60 - LOW	35
E_{C-9}	0.34560 - HIGH	3.60000 - HIGH	0.99 - HIGH	60
E_{C-10}	0.27172 - LOW	1.53521 - LOW	0.99 - HIGH	15

Figure 8.16 illustrates the most interesting case, where the food refresh rate is shown to play a pivotal role in the ecosystem equilibrium states. Table 8.14 summarizes the indicator values for all E_C experiments.

This series has indicated that when the resource availability is low, then the ecosystem cannot reach equilibrium, and all the agents die. Such an outcome is, naturally, highly dependent on the food refresh rate. When food is regularly refreshed, the community survives and expands. This, in fact, is related to the good perception that the agents have of their neighborhood, since there is no need for agents to migrate in search for food. The comparison of experiments E_{C-5} and E_{C-10} supports this claim.

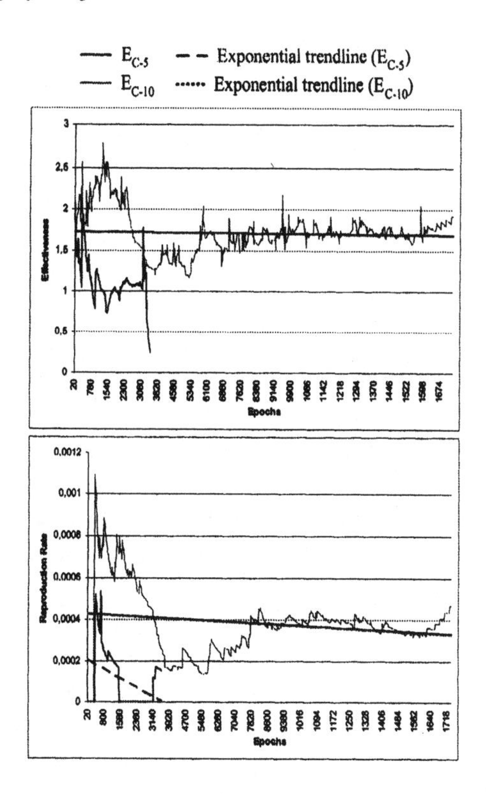

Figure 8.16. The food refresh rate plays a pivotal role in agent survival

Table 8.14. Average indicator values for experiments E_{C-1} to E_{C-10}

Average Indicator Values						
Experiment	Population	e	fcr	$tcr \times 10^{-3}$	ucr	$rr \times 10^{-3}$
E_{C-1}	18.4	1.508	0.122	3.728	0.465	0.265
E_{C-2}	6.2	1.496	0.128	30.465	0.577	0.287
E_{C-3}	14.6	1.555	0.132	9.764	0.464	0.272
E_{C-4}	3.5	1.340	0.138	5.431	0.308	0.109
E_{C-5}	3.4	1.146	0.139	5.534	0.284	0.106
E_{C-6}	2.6	0.751	0.109	26.008	0.445	0
E_{C-7}	7.4	1.389	0.112	35.331	0.689	0.205
E_{C-8}	3.6	0.347	0.093	35.735	0.438	0
E_{C-9}	5.9	1.608	0.142	5.165	0.369	0.289
E_{C-10}	22.7	1.732	0.139	4.319	0.439	0.381

5. Conclusions

Although numerous approaches exist in the research fields of ecosystem simulation and self-organization, Biotope provides an integrated framework for modeling environments, simulating ecosystem behaviors and monitoring system dynamics. It combines primitives drawn from the theories of classifier systems, genetic algorithms, and dispersal distance evolution, in order to produce a reliable, well-built simulation tool. It provides a user-friendly console for tuning and re-tuning the various ecosystems developed. The experimental results have indicated that GAs can, indeed, help in augmenting agent intelligence. Going even further, exploiting the potential of agent communication, gives a boost to facing successfully the problem of simulating unreliable environments. The fact that Biotope has been built upon the methodology described in Chapter 5, provides to the agents of the system the ability of deploying evolutionary DM techniques, in order to augment their intelligence and make it through survival.

Looking at the bigger picture of diffusing knowledge to multi-agent communities, where agents are instantiated and evolve in dynamic, multiparametrical environments with great interdependencies, the way evolutionary DM techniques affect agent intelligence is of particular interest. In contrast to the other two knowledge diffusion levels, where DM techniques are applied on historical data and the extracted knowledge models are embedded into the agents of the each time deployed MAS, in the case of agent communities the focus is given on the appropriate problem modeling. The added-value of evolutionary DM techniques is focalized on the fact that, by following a non-stop recursive process, they succeed in optimizing the fitness function of agents, therefore approximating their ultimate goal.

PART IV

EXTENSIONS...

Chapter 9

AGENT RETRAINING AND DYNAMICAL IMPROVEMENT OF AGENT INTELLIGENCE

A major advantage of the methodology presented in this book is *agent retraining*, the process of revising the knowledge model(s) of agent(s) by re-applying DM techniques. Retraining aspires to improve agent intelligence/performance and can be applied periodically or on a need-based mode. Since there is no point in repeating the DM process on the same dataset, retraining requires either the existence of new application data (Case 1 - Chapter 6), the existence of new behavior data (Case 2 - Chapter 7), or the application of an evolutionary DM technique (Case 3 - Chapter 8).

It is through retraining that we intent to show that certain DM techniques can be used to augment agent intelligence and therefore improve MAS overall performance. In this chapter we provide the formal retraining model for the four DM techniques presented and we appose a number of experimental results.

1. Formal Model

Let us consider a MAS with n agent types that has been developed following the methodology presented in Chapter 5. For each one of the $Q_i, i = 1...n$ agent types of the MAS, DM techniques are applied, in order to produce useful knowledge models KM_o, $o = 1...p$. The corresponding knowledge models are then embedded into the $Q_i(j)$, $j = 1...m$ agents of the system. The MAS is then instantiated.

Once bootstrapped, the MAS engages into *agent monitoring*, one of the most interesting processes in our methodology, which requires each agent to report its actions back to a recording mechanism. This mechanism may vary from an assigned monitoring agent that collects the information (Chapter 6) to a reporting behavior that each agent could carry (Chapter 7), or a monitoring MAS infrastructure (Chapter 8).

In the retraining phase, shown as part of the overall methodology in Figure 9.1, each agent can be retrained individually. The retraining interval is highly dependent on the nature of the MAS application and the role of the agent. The request for retraining can be fired, either by the user/expert that monitors the MAS, or from the agent itself, through an internal evaluation mechanism.

The available datasets include:

i. The initial dataset D_T, including all the D_{IQ} sets (see Section 1 in Chapter 4).

ii. A new non-agent dataset D_{NQ_i}, and

iii. All the datasets $D_{Qi}(j)$, each containing the tuples representing the actions (decisions) taken by the respective agent.

We can also define the agent-type specific dataset, D_{Q_i}, as $D_{Qi} = D_{Qi}(1) \oplus D_{Qi}(2) \oplus ... \oplus D_{Qi}(m)$. The symbol $\oplus$ represents the concatenation of two datasets, an operation that preserves multiple copies of tuples. There are five different options of agent retraining, with respect to the datasets used:

A. $D_{IQ_i} \oplus D_{NQ_i}$
Retrain the agent using the initial dataset along with a new, non-agent dataset D_{NQ_i}.

B. $D_{NQ_i} \oplus D_{Q_i}$
Retrain the agent using a non-agent dataset D_{NQ_i} along with D_{Q_i}, a dataset generated by all the Q_i-type agents of the application. The agents are monitored and their actions are recorded, in order to construct the D_{Q_i} dataset.

C. $D_{IQ_i} \oplus D_{NQ_i} \oplus D_{Q_i}$
Retrain the agent using all the available datasets.

D. $D_{IQ_i} \oplus D_{Q_i}$
Use the initial dataset D_{IQ_i} along with the agent generated data.

E. $D_{IQ_i} \oplus D_{Q_i}(j)$
Use the initial dataset D_{IQ_i} along with $D_{Q_i}(j)$, the generated data of the jth agent of type i.

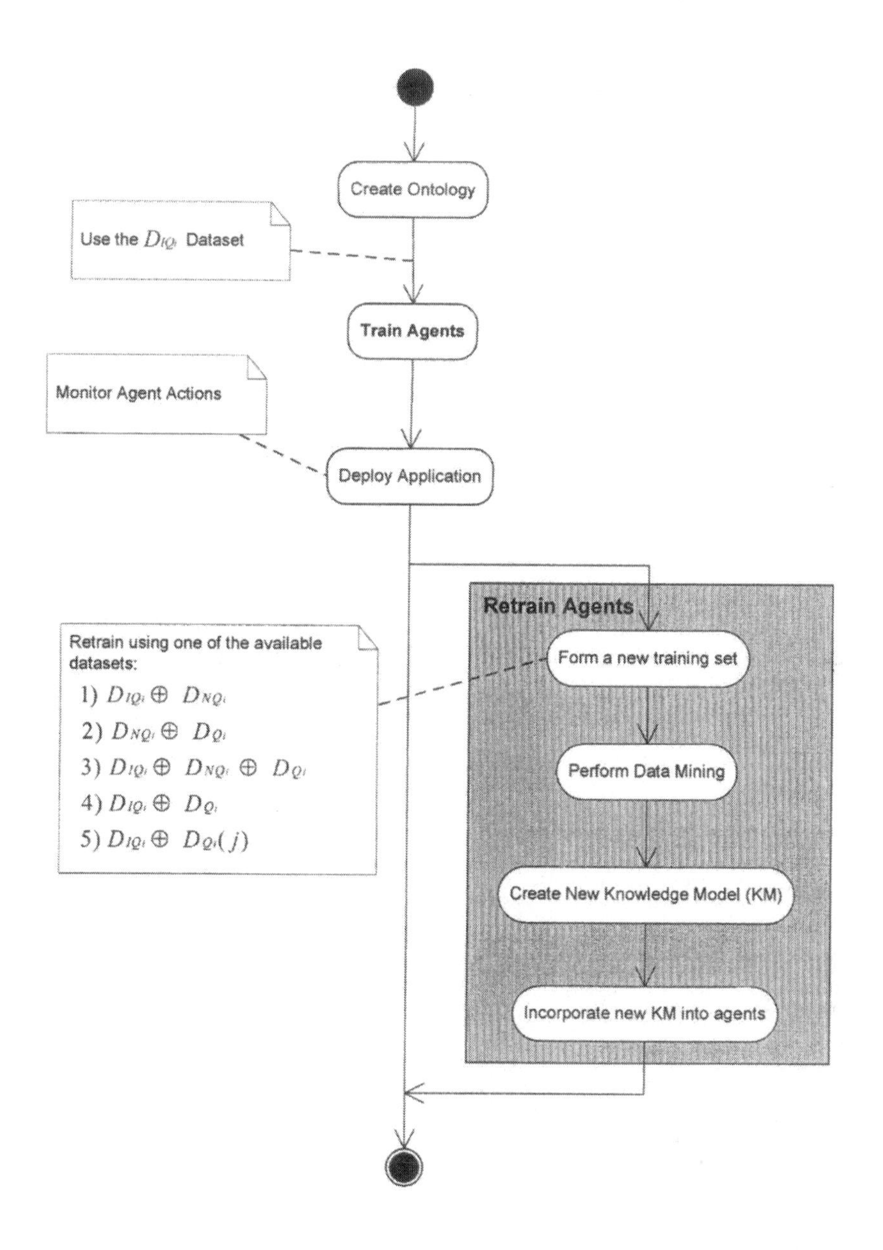

Figure 9.1. Retraining the agents of a MAS

1.1 Different Retraining Approaches

Retraining is performed in order to either increase or refine agent intelligence. By reapplying data mining on a new or more complete dataset, the user expects to derive more accurate patterns and more efficient associations.

The five retraining options defined in the previous section, can be classified into two main approaches: a) the type-oriented, which deals with the improvement of an agent type Q_i and, subsequently, all instances of the same type, and b) the agent-oriented, which focuses on the refinement of an individual agent $Q_i(j)$. The type-oriented approach encompasses options $A - D$, while option E is an agent-oriented approach.

It should also be denoted that we differentiate on the way we define "intelligence improvement", since the presented methodology exploits supervised (classification), unsupervised (clustering and association rule extraction) and evolutionary (genetic algorithms) learning DM techniques. In the case of classification, improvement can be measured by evaluating the knowledge model extracted metrics (mean-square error, accuracy, etc.) and can be achieved only if more data are available (with respect to initial training). In the case of clustering and association rule extraction, retraining is again meaningless without access to additional data, whereas intelligence augmentation is determined by the deployment of external evaluation functions. Finally, in the case of genetic algorithms, no supplementary data are needed, since the technique produces data while trying to optimize the problem it has been employed for.

In the next four sections we discuss agent retraining for each of the four types of DM techniques.

2. Retraining in the Case of Classification Techniques

In this case classification algorithms, such as C4.5 and ID3 discussed in Chapter 2, are used to extract the knowledge models for agents. Although the splitting criteria are different, all classification algorithms are applied in a similar manner. We may focus on the information gain criterion that is employed by the $C4.5$ and $ID3$ algorithms, nevertheless the approach followed can be easily adjusted to other classification algorithms.

2.1 Initial Training

When training takes place, classification is performed on D_{IQ_i}, the initial dataset for the specific agent type. The user may decide to split the dataset into a training and a testing (and/or validation) dataset or to perform n-fold cross-validation. To evaluate the success of the applied classification scheme, a number of statistical measures are calculated, i.e., classification accuracy, mean absolute error, and confusion matrices. If the extracted knowledge model is deemed satisfactory, the user may accept it and store it, for incorporation into the corresponding Q_i-type agents.

2.2 Retraining an Agent Type

In the case of retraining agent-type Q_i, the relevant datasets are D_{IQ_i}, D_{NQ_i} and D_{Q_i}. Retraining option C ($D_{IQ_i} \oplus D_{NQ_i} \oplus D_{Q_i}$) is the most general, containing all the available data for the specific agent type, while options A and D are subsets of option C. They are differentiated, however, since option D is particularly interesting and deserves special attention.

When using datasets D_{IQ_i} and D_{NQ_i}, the user may choose among the different retraining options illustrated in Table 9.1. Notice that each option yields a distinct result.

Table 9.1. Retraining options for $D_{IQ_i} \oplus D_{NQ_i}$

	Dataset		Causality
	D_{IQ_i}	D_{NQ_i}	
Option A–1	Training	Testing	Initial model validation
Option A–2	Testing	Training	Model investigation on Data Independency
Option A–3	Concatenation and Cross-validation		New Knowledge Model discovery

The user decides on which knowledge model to accept, based on its performance. Nevertheless, in the $D_{IQ_i} \oplus D_{NQ_i}$ case, best model performance is usually observed when option $A-3$ is selected. The inductive nature of classification dictates that the use of larger training datasets leads to more efficient knowledge models.

The retraining options when the $D_{NQ_i} \oplus D_{Q_i}$ dataset is selected are illustrated in Table 9.2.

When retraining an agent with the $D_{NQ_i} \oplus D_{Q_i}$ dataset, it is important to notice that the only information we have on the training dataset D_{IQ_i} is indirect, since D_{Q_i} is formatted based on the knowledge model the agents follow, a model inducted by the D_{IQ_i} dataset. This is why the validation of the initial model is indirect. If the D_{NQ_i}-extracted model is similar to the D_{IQ_i}-extracted model, testing accuracy is very high.

Table 9.2. Retraining options for $D_{NQ_i} \oplus D_{Q_i}$

	Dataset		Causality
	D_{NQ_i}	D_{Q_i}	
Option B–1	Training	Testing	Indirect Initial model validation
Option B–2	Concatenation and Cross-validation		New Knowledge Model discovery

The fact that D_{Q_i} is indirectly induced by D_{IQ_i} does not allow testing D_{Q_i} on D_{IQ_i}. Nevertheless, concatenation of the datasets can lead to more efficient and smaller classification models. Since class assignment within D_{Q_i} (the agent decisions) is dependent on the D_{IQ_i}–extracted knowledge model, a "bias" is inserted in the concatenated $D_{IQ_i} \oplus D_{Q_i}$ dataset. Let attribute A_i be the "biased" attribute and C_i the supported class. While recalculating the information gain for the $D_{IQ_i} \oplus D_{Q_i}$ dataset, we observe that the increase of $Info(D)$ is cumulative (Eq. 2.6), while the increase of $Info(D, A_j)$ is proportional (Eq. 2.7) and therefore $Gain(D, A_i)$ is increased. Clearer decisions on the splitting attributes according to the frequency of occurrence of A_i in conjunction to C_i are derived, thus leading to more efficient knowledge models. Table 9.3 illustrates the available retraining options for the corresponding dataset.

Table 9.3. Retraining options for $D_{IQ_i} \oplus D_{Q_i}$

	Dataset		Causality
	D_{NQ_i}	D_{Q_i}	
Option D–1	Concatenation and Cross-validation		More application-efficient Knowledge Model

In the most general case, where all datasets (D_{IQ_i}, D_{NQ_i} and D_{Q_i}) are available, the retraining options are similar to the ones proposed for the already described subsets and similar restrictions apply. Table 9.4 illustrates these options.

2.3 Retraining an Agent Instance

When retraining a specific agent, the user is interested in the refinement of its intelligence in relation to the working environment. Let us assume that we have trained a number of agents that decide on whether a

Table 9.4. Retraining options for $D_{IQ_i} \oplus D_{NQ_i} \oplus D_{Q_i}$

	Dataset			Causality
	D_{IQ_i}	D_{Q_i}	D_{NQ_i}	
Option C–1	Training	Testing	Testing	Initial model validation
Option C–2	Testing	Testing	Training	Model investigation on data independency
Option C–3	Concatenation and Training		Testing	New Knowledge Model (more efficient) validation
Option C–4	Concatenation and Cross-validation			New Knowledge Model discovery

game of tennis should be conducted, according to weather outlook, temperature, humidity and wind conditions (Weather dataset [Blake and Merz, 2000]), and have established these agents in different cities in Greece (Athens, Thessaloniki, Patra, Chania, etc). Although all these agents rely initially on a common knowledge model, weather conditions in Thessaloniki differ from those in Chania enough to justify refined knowledge models.

In this case, we have the options to perform agent-type retraining. By the use of the $D_{IQ_i} \oplus D_{Q_i}$ dataset, it is possible to refine the intelligence of the jth agent of type i. High frequency occurrence of a certain value t_i of attribute A_i (i.e. "High" humidity in Thessaloniki, "Sunny" outlook in Chania) may produce a more "case-specific" knowledge model. In a similar to the $D_{IQ_i} \oplus D_{Q_i}$ manner, it can be seen that an increase of $Info(D, A_j)$ can lead to a different knowledge model, which incorporates instance-specific information.

The analysis of different retraining options in the case of Classification indicates that there exist concrete success metrics that can be used to evaluate the extracted knowledge models and, thus, may ensure the improvement of agent intelligence.

3. Retraining in the Case of Clustering Techniques

In the case of unsupervised learning, training and retraining success cannot be determined quantitatively. A more qualitative approach must be followed, to determine the efficiency of the extracted knowledge model, with respect to the overall goals of the deployed MAS.

3.1 Initial Training

To perform clustering, the user can either split the D_{IQ_i} dataset into a training and a testing subset, or perform a classes-to-clusters evaluation, by testing the extracted clusters with respect to a class attribute defined in D_{IQ_i}. In order to evaluate the success of the clustering scheme, the mean square error and standard deviation of each cluster center are calculated.

3.2 Retraining

Clustering results are in most cases indirectly applied to the deployed MAS. In practice, some kind of an external exploitation function is developed, which somehow fires different agent actions in the case of different clusters. All the available datasets (D_{IQ_i}, D_{NQ_i}, D_{Q_i} and $D_{Q_i}(j)$) can therefore be used for both training and testing for initial model validation, model data dependency investigation and new knowledge model discovery. A larger training dataset and more thorough testing can lead to more accurate clustering. Often retraining can result in the dynamic updating and encapsulation of dataset trends (i.e., in the case of customer segmentation). Retraining $Q_i(j)$ can therefore be defined as a "case-specific" instance of retraining, where data provided by agent j, $D_{Q_i}(j)$, are used for self–improvement.

4. Retraining in the Case of Association Rule Extraction Techniques

4.1 Initial Training

If the user decides to perform ARE on D_{IQ_i}, no training options are provided. Only the algorithm-specific metrics are specified and ARE is performed. In a similar to classification manner, if the extracted knowledge model (clusters, association rules) is favorably evaluated, it is stored and incorporated into the corresponding Q_i-type agents.

4.2 Retraining

The ARE technique does not provide training and testing options. The whole input dataset is used for the extraction of the strongest association rules. Consequently, all available datasets (D_{IQ_i}, D_{NQ_i}, $D_{Q_i}$and $D_{Q_i}(j)$) are concatenated before DM is performed. This unified approach for retraining has a sole goal: to discover the strongest association rules between the items t of D.

The expectation here is that frequently occurring agent actions will reinforce the corresponding rules and possibly improve the knowledge model. In a similar to the clustering case manner, retraining $Q_i(j)$ can be viewed as a "case-specific" instance of retraining.

5. Retraining in the Case of Genetic Algorithms

Through the processes of selection, crossover and mutation (Chapters 2 and 8), the GA mechanism attempts to maximize the case-specific fitness function. Thus, the notion of retraining is inherent in GAs.

6. Experimental Results

In order to prove the added value of agent retraining, a number of experiments on classification, clustering and ARE were conducted. In this section, some representative cases are discussed. The experiments are focused mainly on retraining by the use of the DQ_i and $D_{Q_i}(j)$ datasets and illustrate the enhancement of agent intelligence.

6.1 Intelligent Environmental Monitoring System

The first experiment was performed for the O3RTAA System (Figure 9.2), an agent-based intelligent environmental monitoring system developed for assessing ambient air-quality [Athanasiadis and Mitkas, 2004]. A community of software agents is assigned to monitor and validate multi-sensor data, to assess air-quality, and, finally, to fire alarms to appropriate recipients, when needed. Data mining techniques have been used for adding data-driven, customized intelligence into agents with successful results.

In this work we focused on the Diagnosis Agent Type. Agents of this type are responsible for monitoring various air quality attributes including pollutants' emissions and meteorological attributes. Each one of the Diagnosis Agent instances is assigned to monitor one attribute through the corresponding field sensor. In the case of sensor breakdown, Diagnosis Agents take control and perform an estimation of the missing sensor values using a data-driven reasoning engine, which exploits DM techniques.

One of the Diagnosis Agents is responsible for estimating missing ozone measurement values. This task is accomplished using a predictive model comprised of the predictors and the response. For the estimation of missing ozone values the predictors are the current values measured by the rest of the sensors, while the response is the level of the missing value (Low, Medium, or High). In this way, the problem has been formed as a classification task.

For training and retraining the Ozone Diagnosis Agent we used a dataset, labeled C2ONDA01 and supplied by CEAM, the Centro de Es-

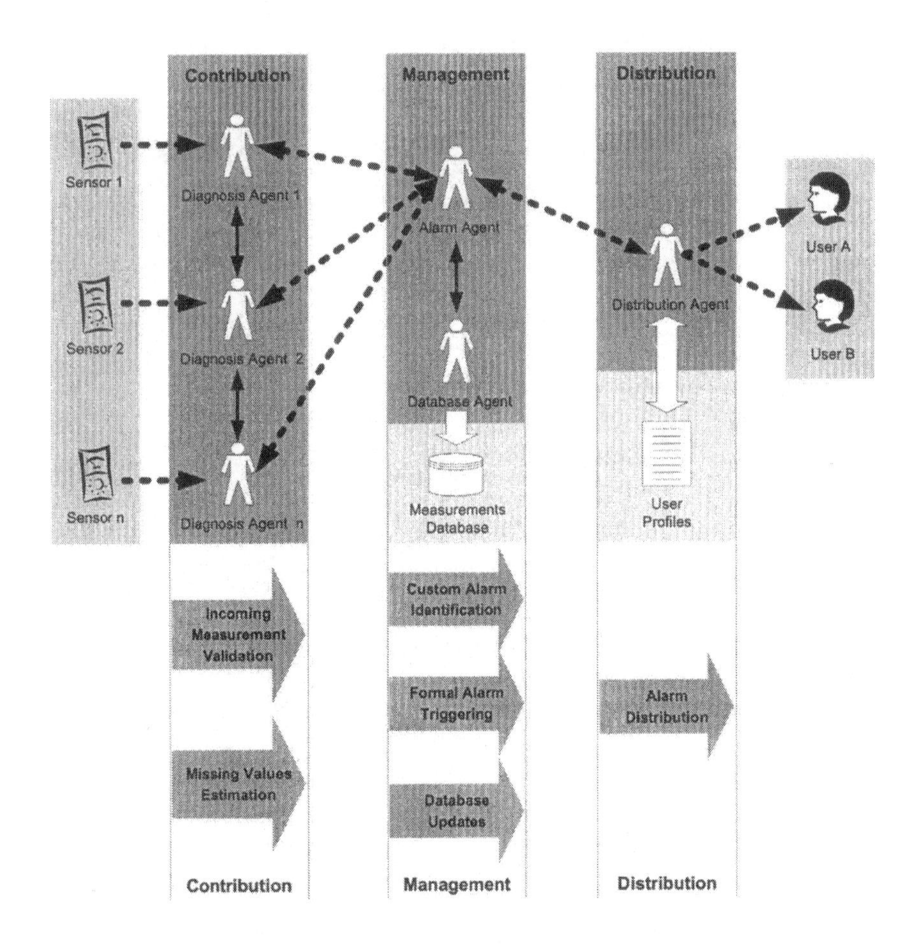

Figure 9.2. The O3RTAA system architecture

tudios del Mediterraneo in Valencia, Spain. The dataset contained data from a meteorological station in the district of Valencia, Spain. Several meteorological attributes and air-pollutant values were recorded on a quarter-hourly basis during the year 2001. There were approximately 35,000 records, with ten attributes per record plus the class attribute. The dataset was split into three subsets: one subset for initial training (D_{IQ_i}), a second subset for agent testing (D_{Q_i}) and another subset for validation (D_{Val}) containing around 40%, 35% and 25% of the data, respectively.

The initial training of the Diagnosis Agent was conducted using Quinlan's C4.5 algorithm for decision tree induction. The algorithm was applied on the D_{IQ_i} subset. The induced decision tree was embedded in the Diagnosis Agent and the agent used it for deciding on the records of the D_{Q_i} subset. Agent decisions along with the initial application data were used for retraining the Diagnosis Agent (Option D: $D_{IQ_i} \oplus D_{Q_i}$). Finally, the Diagnosis Agent with the updated decision tree was used for deciding on the cases of the last subset (D_{Val}).

The retrained Diagnosis Agent performed much better compared to the initial training model, as shown in Table 9.5. The use of agent decisions included in D_{Q_i} has enhanced the Diagnosis Agent performance on the D_{Val} subset by 3.65%.

Table 9.5. Classification accuracies for the Diagnosis Agent

	Dataset		
	D_{IQ_i}	D_{Q_i}	D_{Val}
Number of instances	11641	10000	7414
Initial training	Used	73.58%	71.89%
Retraining	Used		74.66%

6.2 Speech Recognition Agents

This experiment was based on the "vowel" dataset of the UCI repository [Blake and Merz, 2000]. The problem in this case was to recognize a vowel spoken by an arbitrary speaker. This dataset comprises ten continuous primary features (derived from spectral data) and two discrete contextual features (the speaker's identity and sex) and contains records for 15 speakers. The observations fall into eleven classes (eleven different vowels).

The vowel problem was assigned to an agent community to solve. Two agents $Q_i(1)$ and $Q_i(2)$ were deployed to recognize vowels. Although of the same type, the two agents operated in different environments. This is why the dataset was split in the following way: The data of the first nine speakers (D_{IQ_i}) were used as a common training set for both $Q_i(1)$ and $Q_i(2)$. The records for the next two speakers were assigned to $Q_i(1)$ and those of the last two speakers were assigned to $Q_i(2)$.

The procedure followed was to evaluate the retraining performance of each on of the agents (Option E: $D_{IQ_i} \oplus D_{Q_i}(j)$). After initial training with D_{IQ_i}, each of the $Q_i(1)$ and $Q_i(2)$ was tested on one of the two assigned speakers, while the second speaker was used for the evaluation of the retraining phase. Quinlan's C4.5 algorithm was again applied. The classification accuracy, which is similar to that reported by P.D. Turney [Turney, 2002], is illustrated in Table 9.6.

Table 9.6. Speech Recognition Agents Classification accuracy

	$\mathbf{Q}_i(1)$			$\mathbf{Q}_i(2)$		
	D_{IQ_i}	$D_{Q_i}(1)$	$D_{Val}(1)$	D_{IQ_i}	$D_{Q_i}(2)$	$D_{Val}(2)$
Number of speakers	9	1	1	9	1	1
Initial training	Used	53.03%	46.97%	Used	33.33%	28.78%
Retraining	Used		56.06%	Used		43.93%

It is obvious in this case that retraining using $D_{Q_i}(j)$ leads to considerable enhancement of the agents' ability to decide correctly. The decision models that are induced after the retraining procedure outperformed the validation speakers. The improvement in terms of classification accuracy was about 36% in average.

6.3 The Iris Recommendation Agent

In order to investigate retraining in the case of clustering, we used the Iris UCI Dataset [Blake and Merz, 2000], a dataset widely used in pattern recognition literature. It has four numeric attributes describing the iris plant and one nominal attribute describing its class. The 150 records of the set were split into two subsets: one subset (75%) for initial training (D_{IQ_i}) and a second subset (25%) for agent testing (D_{Q_i}). Classes-to-clusters evaluation was performed on D_{IQ_i} and $D_{IQ_i} \oplus D_{Q_i}$ (Option D) and the performance of the resulted clusters was compared on the number of correctly classified instances of the dataset (Table 9.7).

Table 9.7. The Iris Recommendation Agent success

	$\mathbf{Q}_i$		
	D_{IQ_i}	D_{Q_i}	**Correctly Classified**
Number of records	113	37	
Initial training	Used	-	83.19%
Retraining	Used		88.67%

Again, retraining with the $D_{IQ_i} \oplus D_{Q_i}$ dataset leads to the improvement of clustering results.

7. Conclusions

The work presented explains how the concept of retraining, the iterative process of "recalling" an agent for posterior training, is formulated. Through this procedure, where DM is performed on new datasets (D_{NQ_i}, D_{Q_i} and $D_{Q_i}(j)$), refined knowledge is extracted and dynamically embedded into the agents.

For the first two levels of knowledge diffusion, retraining is a process that can (although not advised) be skipped. The three experiments discussed here illustrate that there can be improvement in the efficiency of knowledge models, therefore leading indirectly to the improvement of agent reasoning.

In the third level of knowledge diffusion, however, retraining constitutes a prominent part of the agents' reasoning mechanism. Whenever an agent acts, an evaluation mechanism is invoked, so as to reward/punish the specific rule that urged the agent to act that way.

Based on this preliminary research work we strongly believe that retraining should be applied for all types of agents and can, indeed, lead to intelligence augmentation.

Chapter 10

AREAS OF APPLICATION & FUTURE DIRECTIONS

1. Areas of Application

We have mentioned repeatedly in this book that the agent paradigm can be adopted by a great variety of application domains, since agents constitute the software building blocks of a new generation. Nevertheless, for a fruitful coupling of agent technology with data mining, either historical data (of any type) must be available, or the knowledge mechanisms of agents must allow self-organization.

In Chapters 6 to 8 we have discussed in considerable detail three representative application domains. Here, we briefly outline some additional areas that exhibit the above characteristics and, thus, are good candidates for agent training.

1.1 Environmental Monitoring Information Systems

The field of Environmental Informatics (also known as Enviromatics) encompasses a wide range of applications from expert systems for environmental impact assessment, to intricate environmental simulators. National and international legislation increasingly dictates that the public must be informed of environmental changes and alerted, in case of emergencies and hazards. This requirement has fueled the development of a whole class of applications called Environmental Monitoring Information Systems (EMIS). The primary function of EMIS is the continuous monitoring of several environmental indicators in an effort to produce sound and validated information. The architectures and functionalities of EMIS vary from naive configurations, focused on data collection and projection, to elaborate decision-support frameworks dealing with phenomena surveillance, data storage and manipulation, knowledge discovery and diffusion to the end users (government agencies, non-governmental organizations, and citizens). The presented synergy of AT and DM can provide to EMIS efficient solutions for the monitoring, management and distribution of environmental changes, while eliminating the time-overhead that often exists between data producers and data consumers.

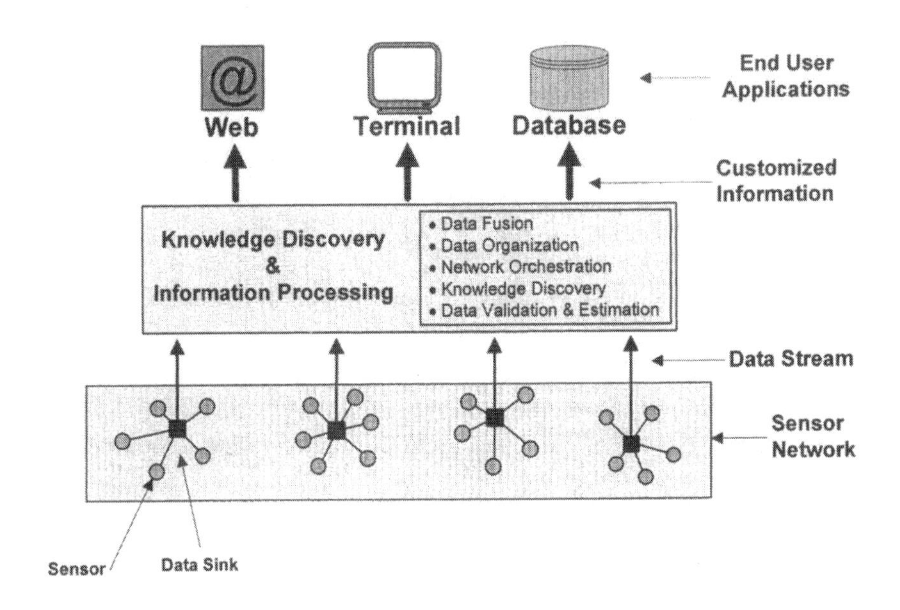

Figure 10.1. The generalized EMIS architecture

A generalized architecture for EMIS is illustrated in Figure 10.1. The framework is situated upon a sensor network and interweaves multiple data streams in order to supply the end-user applications with pre-processed, ready-to-use information. These end-user applications could be database systems, domain-related software applications, or web-services.

An environmental monitoring system that exploits the AT–DM synergy exhibits two key improvements, with respect to traditional (even agent-based) architectures: a) the ability to deal with data uncertainty problems and b) to provide a proactive communication mechanism for connecting sensor networks with user applications. The former can be realized by adding intelligent features, while the latter relies on the framework's adaptability. The main functionalities of such a framework can be summarized to the following:

1. Data collection and validation
2. Data management and pre-processing
3. Knowledge discovery
4. Network orchestration
5. Proactive communication
6. Information Propagation

1.2 Agent Bidding and Auctioning

Software agents that participate in electronic business transactions in an effort to increase revenue for humans introduce a new form of automatic negotiations. These agents enjoy a large degree of autonomy because, despite being computer programs, they actually undertake the responsibility of deal making on behalf of humans. In this way, agents can mitigate the difficult task of having to deliberate over offering the best price for purchasing a particular item. Even though humans seem to negotiate using complex procedures, it is not feasible for them to monitor and understand the attributes of a sequence of negotiations in a large competitive environment, such as the one that hosts electronic auctions (e-auctions). Besides, agent superiority in terms of monitoring and "remembering", agents are also able to follow a specific course of action towards their effort to increase their profit without being diverted by emotional influence as humans do.

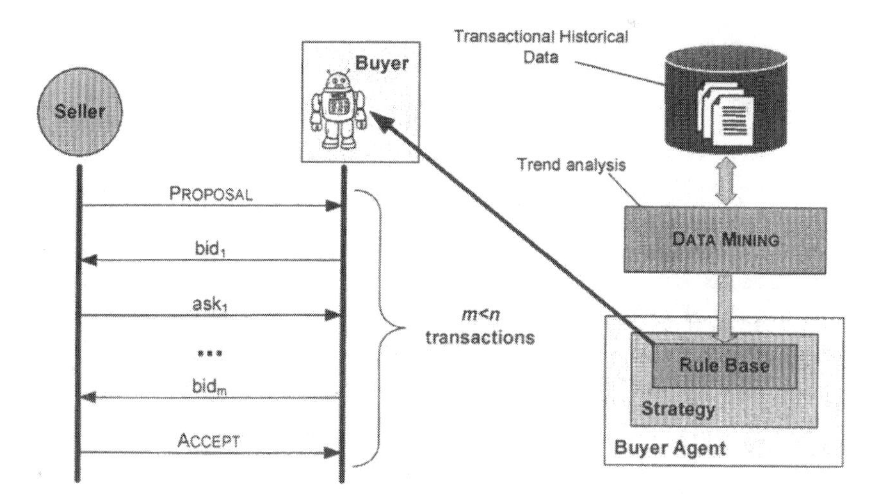

Figure 10.2. Improving the behavior of biding agents

An interesting issue concerning agents in e-commerce is the creation of both rational and efficient agent behaviors, to enable reliable agent-mediated transactions. In fact, through the presented methodology the improvement of agent behaviors in auctions is feasible. Data mining can be performed on available historical data describing the bidding flow and the results can be used to improve the bidding mechanism of agents for e-auctions. Appropriate analysis of the data produced as an auction progresses (historical data) can lead to more accurate short-term forecasting [Kehagias et al., 2005]. By the use of trend analysis techniques, an agent can comprehend the bidding policy of its rivals, and, thus, re-adjust its own in order to yield higher profits for buyers. In addition, the number of negotiations between interested parties is reduced (m instead of n, $m < n$), since accurate forecasting implies more efficient bidding (Figure 10.2).

1.3 Enhanced Software Processing

Multi-agent systems can also be exploited for the improvement of the software development process. Software houses often develop frameworks for building end-user applications, following a standard methodology. Such frameworks interconnect different software modules and combine different software processes into workflows, in order to produce the desired outcome (Figure 10.3). Apart from the scheduling and planning, where agents collaborate and negotiate to reach the optimal solution, data mining techniques can be applied on workflow graphs, in order to discover correlations between certain workflow traverses. In addition, DM techniques can be applied to solve any kind of application-level problems, i.e., decision making capabilities.

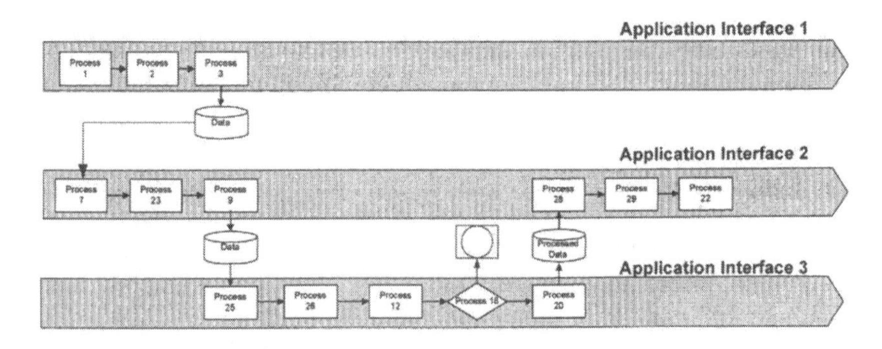

Figure 10.3. A software workflow process

The core advantages of adopting the AT–DM synergy in such frameworks would be:

- Accelerated development processes
- Enhanced variety of applications offered to customers
- Enhanced customer benefit
- Reduction in application development costs

2. Advanced AT–DM Symbiosis Architectures

The presented methodology can be applied to MAS by the use of standard AOSE tools (see Chapter 4). Nevertheless, elaborate architectures can be implemented, taking seriously into account issues related to scalability, portability, distributivity, and semantic-awareness. Two such advanced architectures are presented next.

2.1 Distributed Agent Training Architectures

Drawing from the primitives of Virtual Networks and Cluster Organizations, one could imagine a collaborative distributed framework that would promise to deliver truly "smart" solutions in the area of application services provision. Such a framework would exploit the AT–DM synergy, in order to provide services in various application domains, i.e., production planning, advanced Business-to-Business (B2B) transactions, and real-time systems management, overcoming the obstacles of increased cost and expertise requirements.

The main goal of such an effort would be to define all the requirements and develop the tools that create, train and return agents on behalf of individual users, specialized on their demands and requests. Specific objectives would include:

1. The integration of Service Provision by developing an intelligent and user-decentralized knowledge platform.
2. The reduction of cost for users that desire to embed agent intelligence in their organization strategy.
3. The development of a generic model for the "meeting" of agents and the knowledge exchange.
4. The creation of a reusable and configurable platform for supporting the provision of added-value services among the facilitating virtual organizations and the facilitated user companies.

A possible scenario would involve a software developer, who, having installed a number of MAS applications for its customers, chooses to maintain a common knowledge repository augmented with facilities for agent monitoring and retraining (see Figure 10.4). The company, then, may add to the usual support contract the option to periodically upgrade a customer's agents by 'recalling' them for retraining. In reality, an agent scheduled for upgrading would receive a new knowledge model to replace the current one. As we have already mentioned in the previous chapter, retraining highly depends on the availability of new data. A customer may opt to restrict agent tracking and data collection in house or to contribute application- and agent-specific data to the common knowledge base.

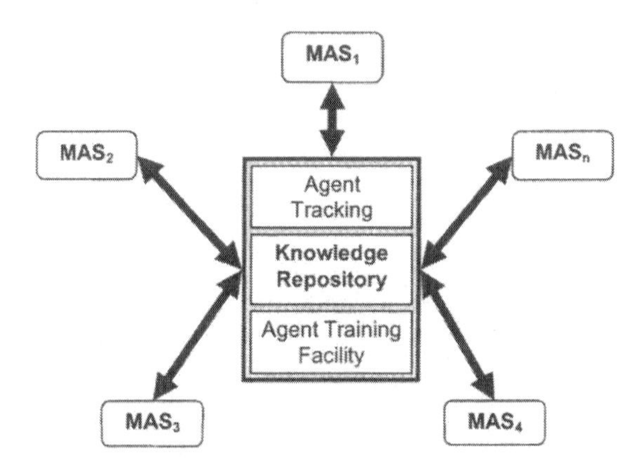

Figure 10.4. A common knowledge repository and training facility

2.2 Semantically-Aware Grid Architectures

An even more advanced architecture is illustrated in Figure 10.5 and aims to develop an open, scalable framework that will enable the semantic integration of heterogeneous information resources. Such a framework is expected to enhance knowledge sharing between services and applications in the semantic grid.

By combining agent technologies with a variety of techniques, including data mining, case-based reasoning (CBR), summarization and annotation, one can achieve automation of the knowledge lifecycle, enabling acquisition, classification, managing and sharing of knowledge. The results of the application of this system include knowledge fusion and semantic interoperability over applications, which cross organizational boundaries and application domains.

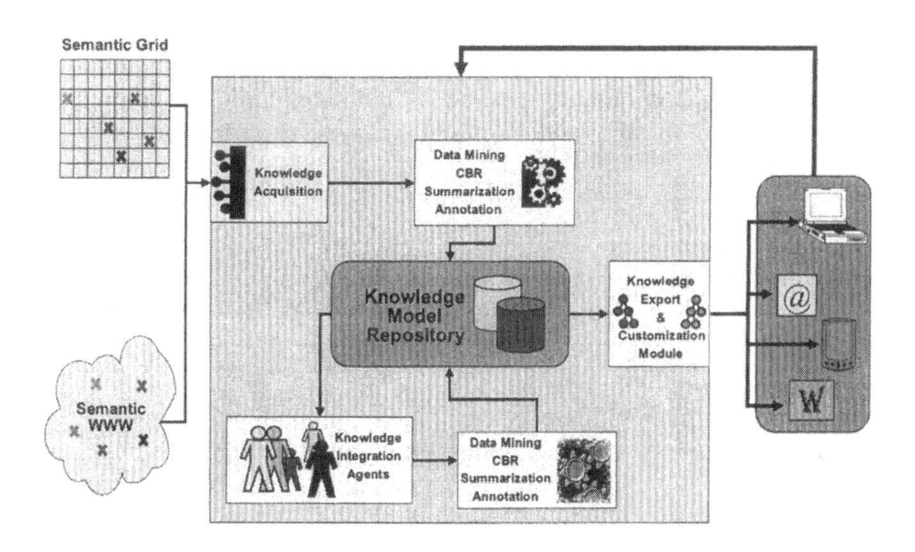

Figure 10.5. A semantically-aware framework for increasing agent intelligence

3. Summary and Conclusions

The use of agent technology leads to systems characterized by both autonomy and a distribution of tasks and control [Sycara et al., 1998], while the utilization of data-mined knowledge offers the capability of deploying "*Interesting (non-trivial, implicit, previously unknown, and potentially useful) information or patterns from data in large databases*" [Fayyad et al., 1996].

There is a plethora of systems where agent intelligence is based on simple rules, leveraging the cost of development, tractability and maintenance [Jennings and Wooldridge, 1998]. In dynamic and complicated environments, however, simple expert engines prove insufficient. A series of complex and dynamic rules need to be applied, leading to conflicts and high human involvement, which results to the depletion of rationality of human experts. Moreover, automatic adaptability is not an issue, since it is exhausted to probabilistic inference [Caglayan and Harrison, 1997]. In the presented work, we argue that the synergy of data mining technology with agent-based applications offers a strong ***autonomous*** and ***adaptable*** framework.

Within the context of this book we have presented a methodology that takes all limitations into account and provides the ability to *dynamically* embed DM-extracted knowledge to agents and multi-agent systems. Three different types of knowledge are identified, with respect to the data mining techniques applied, leading to three distinct ways of knowledge diffusion to the MAS architecture. These are: a) knowledge extracted by use of DM techniques on the application level of a MAS (macroscopic view), b) knowledge extracted by the use of DM techniques on the behavior level of a MAS, (microscopic view), and c) knowledge extracted by the use of evolutionary DM techniques on evolutionary agent communities.

In this book we have considered the applicability potential of the four dominant DM techniques on MAS: classification for categorization and prediction, clustering for grouping, association rule extraction for correlation discovery, and genetic algorithms for self-organization. In order to test and demonstrate the concepts of agent *training* and *retraining*, we have implemented Data Miner (see Chapter 5). For each one of the knowledge types identified, a representative demonstrator was built and is presented (Chapters 6-8). The process of revising, in order to improve the knowledge model(s) of agent(s), by re-applying DM techniques on new, up-to-date data was discussed in Chapter 9. Finally, representative application domains and extended architectures for our methodology were provided in the first section of this Chapter.

The advantages of the presented methodology that couples data mining techniques with agent technology primitives can be summarized to the following:

- The combination of autonomy (MAS) and knowledge (DM) provides adaptable systems. The process itself rejects profiles, rules and patterns not in use, while it adopts others that are frequently encountered.

- The rigidity and lack of exploration of deductive reasoning systems is overcome (see Chapter 4). Rules are no longer hard-coded into systems and their modification is only a matter of retraining.

- Techniques such as association rule extraction have no equivalent in expert systems. These techniques now provide the agents with the ability of probing and searching.

- Real-world databases often contain missing, erroneous data and/or outliers. Through clustering, noisy logs are assimilated and become a part of a greater group, smoothing down differences (i.e. IPRA), while outliers are detected and rejected. Through classification, ambiguous data records can be validated and missing data records can be estimated. Rule-based systems cannot handle such data efficiently without increasing their knowledge-base and therefore their maintenance cost.

- The presented approach favors the combination of inductive and deductive reasoning models. In some of the demonstrators presented, there were agents deploying deductive reasoning models, ensuring therefore system soundness. Nevertheless, these agents decide on data already preprocessed by inductive agents. That way, the dynamic nature of the application domains is satisfied, while set of deductive results (knowledge-bases of deductive agents) become more compressed and robust.

- Even though the patterns and rules generated through data mining cannot be defined as *sound*, there are metrics deployed to evaluate the performance of the algorithms. Total mean square error (clustering), support-confidence (association rules), classifier accuracy (classification), classifier evaluation (genetic algorithms) among others, are employed for evaluating the knowledge models extracted. Our approach takes under serious consideration the need for knowledge model evaluation and provides through the Data Miner a series of functionalities for visualization, model testing and model comprehension.

- Usually DM tools are introduced to enterprises as components-off-the-self (COTS). These tools are used by human-experts to examine their corporate or environmental databases and then develop strategies and take decisions. This procedure often proves time-consuming and inefficient. By exploiting concurrency and multiple instantiation of agent types (cloning capabilities) of MAS systems, and by applying data mining techniques for embedding intelligent reasoning into them, useful recommendations can be much faster diffused while parallelism can be applied to non-related tasks pushing system performance even higher.

It is, thus, evident that the coupling of agent technology and data mining can provide substantial benefits and overcome a number of barriers, usually met in expert systems, hindering the unflustered diffusion of knowledge. The methodology can provide added-value to agents and multi-agent systems and should be pursued.

4. Open Issues and Future Directions

This book is mainly focused on the specification of a concrete methodology that can be adopted to build agents and multi-agent systems with the capability of dynamically incorporating and exploiting DM-extracted knowledge. It has already been noted that the application domains of the constituent technologies are vast, therefore no extensive analysis of all the related issues would have been feasible. Open issues and questions still remain; they always do. The most interesting of them are:

1. *The development of an analytical methodology for agent retraining* The analysis is Chapter 9 is compendious, only introducing and formalizing the concepts related to the revising of knowledge models. In order for an in-depth analysis on retraining, the specification of metrics and scales capable of evaluating agent intelligence is essential, as well as the specification of the appropriate retraining conditions and intervals.

2. *The development of a methodology for evaluating MAS efficiency* Going through related bibliography, one can identify that no established generic methodology exists for evaluating MAS performance exists. Thus, an interesting elaboration of the presented research work would be the foundation of a robust theoretical basis for evaluating the efficiency of agents and multi-agent systems, as well as the development of a proper framework that would incorporate this basis and benchmark real-world, for real-life agent applications.

GLOSSARY

AA Agent Academy

ACL Agent Communication Language

AFLIE Adaptive Fuzzy Logic Inference Engine

AI Artificial Intelligence

AOSE Agent-Oriented Software Engineering

API Application Programming Interface

ARE Association Rule Extraction

AVP Average Visit Percentage

CAS Complex Adaptive Systems

CBR Case-Based Reasoning

CF Clustering Feature

CL Classification

CLS Clustering

COA Customer Order Agent

COTS Component-off-the-self

CPIA Customer Profile Order Agent

CV Corresponding Value

DL Deductive Logic

DM Data Mining

DS Decision Support

DT Decision Tree

DW Data Warehouse

ear Energy Availability Rate

EMIS Environmental Monitoring Information Systems

elr Energy Loss Rate

ERP Enterprise Resource Planning

ERPA Enterprise Resource Planning Agent

eu Energy Unit

eur Energy Uptake Rate

fcr Food Consumption Rate

FIS Fuzzy Inference System

FR Fuzzy Rule

FV Fuzzy Value

GA Genetic Algorithms

IL Inductive Logic

IPIA Inventory Profile Identification Agent
IPRA Intelligent Policy Recommendation Agent
IR Information Retrieval
IRF Intelligent Recommendation Framework
IT Information Technology
JADE Java Agent Development Framework
JAFMAS Java-based Agent Framework for Multi-Agent Systems
JATLite Java Agent Template, Lite
KDD Knowledge Discovery in Databases
KIF Knowledge Interchange Format
KM Knowledge Model
KQML Knowledge and Querying Manipulation Language
KSE Knowledge Sharing Effort
MAS Multi-Agent System
NAR Next Action Recommendation Score
OV Output Values
PSSR Prediction System Success Rate
RA Recommendation Agent
Rec Recommendation Score
RMI Remote Method Invocation
rr Reproduction Rate
RS Recommendation Set
SA Software Agent
SC Supply Chain
SE Software Engineering
SME Small-Medium Enterprise
SPIA Supplier Profile Order Agent
tcr Trap Collision Rate
usr Unknown situation Rate
WAVP Weighted Average Visit Percentage

References

Ackley, David H. and Littman, Michael S. (1990). Learning from natural selection in an artificial environment. In *IEEE International Joint Conference on Neural Networks*, volume Theory Track, Neural and Cognitive Sciences Track, pages 189–193. IEEE (Erlbaum Assoc. Publishers).

Adriaans, Pieter and Zantinge, Dolf (1996). *Data Mining*. Addison-Wesley, Reading, Massachusetts.

Agarwal, Ramesh C., Aggarwal, Charu C., and Prasad, V. V. V. (2001). A tree projection algorithm for generation of frequent item sets. *Journal of Parallel and Distributed Computing*, 61(3):350–371.

Agent Working Group (2000). Agent technology green paper. Technical report, Object Management Group.

Agrawal, Rakesh and Srikant, Ramakrishnan (1994). Fast algorithms for mining association rules. In Bocca, Jorge B., Jarke, Matthias, and Zaniolo, Carlo, editors, *Proc. 20th Int. Conf. Very Large Data Bases, VLDB*, pages 487–499. Morgan Kaufmann.

Amir, Amihood, Feldman, Ronen, and Kashi, Reuven (1999). A new and versatile method for association generation. *Information Systems*, 22:333–347.

Arthur, B. W. (1994). Inductive reasoning and bounded rationality. *American Economic Review*, 84:406–411.

Athanasiadis, I. N. and Mitkas, P. A. (2004). An agent-based intelligent environmental monitoring system. *Management of Environmental Quality*, 15:229–237.

Banerjee, Arindam and Ghosh, Joydeep (2001). Clickstream clustering using weighted longest common subsequences.

Barbará, Daniel, DuMouchel, William, Faloutsos, Faloutsos, Haas, Peter J., Hellerstein, Joseph M., Ioannidis, Yannis E., Jagadish, H. V., Johnson, Johnson, Ng, Raymond T., Poosala, Viswanath, Ross, Kenneth A., and Sevcik, Kenneth C. (1997). The New Jersey data reduction report. *IEEE Data Engineering Bulletin: Special Issue on Data Reduction Techniques*, 20(4):3–45.

Bellifemine, Fabio, Poggi, Agostino, and Rimassa, Rimassa (2001). Developing multi-agent systems with JADE. *Lecture Notes in Computer Science*, 1986:89–101.

Bigus, J. P. (1996). *Data Mining with Neural Networks Solving Business Problems from Application Development to Decision Support*. Mc Graw-Hill.

Birmingham, W.P. (2004). Essential issues in distributed computational systems.

Blake, C. L. and Merz, C. J. (2000). UCI repository of machine learning databases.

Booker, L., Goldberg, D. E., and Holland, J. H. (1989). Classifier systems and genetic algorithms. *Artificial Intelligence*, 40:235–282.

Bossel, H. (1977). *Orientors of Nonroutine Behavior*, pages 227–265. Verlag.

Bousquet, F., Cambier, C., and Morand, P. (1994). Distributed artificial intelligence and object-oriented modelling of a fishery. *Mathematical and Computer Modelling*, 20(8):97–107.

Bretthorst, G. L. (1994). An introduction to model selection using probability theory as logic. In Heidbreder, G., editor, *Maximum Entropy and Bayesian Methods*. Kluwer Academic.

Caglayan, Alper K. and Harrison, Colin G. (1997). Agent sourcebook.

Cardelli, Luca (1995). A language with distributed scope. In *Conference Record of POPL '95: 22nd Annual ACM SIGPLAN-SIGACT Symposium on Principles of Programming Languages, San Francisco, Calif.*, pages 286–297, New York, NY.

Carlsson, C. and Turban, E. (2002). Dss: directions for the next decade. *Decision Support Systems*, 33:105–110.

Caswell, Hal (1989). *Matrix Population Models: Construction, Analysis, and Interpretation.* Sinauer Associates, Inc.

Chang, Daniel T. and Lange, Danny B. (1996). Mobile agent: A new paradigm for distributed object computing on the www. In OOPSLA'96 Workshop: Toward the Integration of WWW and Distributed Object Technology.

Chen, Ming-Syan, Han, Jiawei, and Yu, Philip S. (1996). Data mining: an overview from a database perspective. *Ieee Trans. On Knowledge And Data Engineering*, 8:866–883.

Chen, Z. (1999). *Computational Intelligence for Decision Support.* CRC Press.

Choy, K. L., Lee, W. B., and Lo, V. (2002). Development of a case based intelligent customer-supplier relationship management system. *Expert Systems with Applications*, 23:281–297.

Choy, K. L., Lee, W. B., and Lo, V. (2003). Design of an intelligent supplier relationship management system: a hybrid case based neural network approach. *Expert Systems with Applications*, 24:225–237.

Colin, Andrew (1996). Algorithm alley: Building decision trees with the ID3 algorithm. *Dr. Dobb's Journal of Software Tools*, 21(6):107–109.

Cooley, Robert, Mobasher, Bamshad, and Srivastava, Jaideep (1999). Data preparation for mining world wide web browsing patterns. *Knowledge and Information Systems*, 1(1):5–32.

Crist, T. O. and Haefner, J. W. (1994). Spatial model of movement and foraging in harvester ants (pogonomyrmex) (ii): The roles of memory and communication. *Journal of Theoretical Biology*, 166:315–323.

Data Mining Group (2001). Predictive model markup language specifications (pmml), ver. 2.0. Technical report, The DMG Consortium.

Davenport, T. H. (2000). The future of enterprise system-enabled organizations. *Information Systems Frontiers*, 2:163–180.

Dean, J. (1998). Animats and what they can tell us. *Trends in Cognitive Sciences*, 2:60–67.

DeAngelis, Donald L. and Gross, Louis J., editors (1992). *Individual-based models and approaches in ecology: populations, communities, and ecosystems.* Routledge, Chapman and Hall, New York, NY.

Degen, Wolfgang, Heller, Barbara, Herre, Heinrich, and Smith, Barry (2001). GOL: A general ontology language. In Welty, Christopher, Barry Smith, editor, *Proceedings of the 2nd International Conference on Formal Ontology in Information Systems*, pages 34–45, New York. ACM Press.

Durrett, Rick (1999). Stochastic spatial models. *SIAM Review*, 41(4):677–718.

Eiter, Thomas and Mascardi, Viviana (2002). Comparing environments for developing software agents. *AI Communications*, 15(4):169–197.

Epstein, J. M. and Axtell, R. L. (1996). *Growing Artificial Societies: Social Science from the Bottom Up*. The MIT Press.

Ester, Martin, Kriegel, Hans-Peter, Sander, Jorg, and Xu, Xiaowei (1996). A density-based algorithm for discovering clusters in large spatial databases with noise. In Simoudis, Evangelos, Han, Jiawei, and Fayyad, Usama, editors, *Second International Conference on Knowledge Discovery and Data Mining*, pages 226–231, Portland, Oregon. AAAI Press.

Fayyad, Usama (1996). Mining databases: Towards algorithms for knowledge discovery. *Bulletin of the Technical Committee on Data Engineering*, 21:39–48.

Fayyad, Usama M., Piatetsky-Shapiro, Gregory, and Smyth, Padhraic (1996). Knowledge discovery and data mining: Towards a unifying framework. In *Knowledge Discovery and Data Mining*, pages 82–88.

Ferber, J. (1999). *Multi-Agents Systems - An Introduction to Distributed Artificial Intelligence*. Addison-Wesley.

Fernandes, A. A. A. (2000). Combining inductive and deductive inference in knowledge management tasks. In *Proceedings of the 11th International Workshop on Database and Expert Systems Applications*, pages 1109–1114. IEEE Computer Society.

Finin, Tim, Labrou, Yannis, and Mayfield, James (1997). KQML as an agent communication language. In Bradshaw, Jeffrey M., editor, *Software Agents*, chapter 14, pages 291–316. AAAI Press / The MIT Press.

Fox, G., Pierce, M., Gannon, D., and Thomas, M. (2002). Overview of grid computing environments. Technical report, Global Grid Forum. GGF Document Series.

Frederiksson, M. and Gustavsson, R. (2001). A methodological perspective on engineering of agent societies. In Omiccini, A., Zambonelli, F., and Tolksdorf, R., editors, *Engineering societies in the agents world*, volume 2203, pages 1–16. Springer-Verlag.

Freitas, A. A. (1999). On rule interestingness measures. *Knowledge-Based Systems*, 12:309–315.

Friedman, Jerome H. (1977). A recursive partitioning decision rule for nonparametric classifiers. *IEEE Transaction on Computer*, 26:404–408.

Friedman-Hill, Ernest J. (1998). *Jess, The Java Expert System Shell*. Sandia National Laboratories, Livermore, CA, USA.

Galitsky, B. and Pampapathi, R. (2003). Deductive and inductive reasoning for processing the claims of unsatisfied customers. In *Proceedings of the 16th IEA/AIE Conference*, pages 21–30. Springer-Verlag.

Ganti, Venkatesh, Gehrke, Johannes, and Ramakrishnan, Raghu (1999). Mining very large databases. *Computer*, 32(8):38–45.

Gasser, L. (1991). Social conceptions of knowledge and action: Dai foundations and open systems semantics. *Artificial Intelligence*, 47:107–138.

Genesereth, M. R. and Ketchpel, S. (1994). Software agents. *Communications of the ACM*, 37:48–53.

Genesereth, Michael R. and Fikes, Richard E. (1992). Knowledge interchange format, version 0.3. Technical report, Knowledge Systems Laboratory, Stanford University, Stanford, California.

Goldberg, Adele and Robson, David (1983). *Smalltalk-80: The Language and its Implementation*, pages 674–681. Addison-Wesley.

Goldberg, D. (1989). *Genetic Algorithms in Search, Optimization, and Machine Learning*, pages 1–24. Addison-Wesley.

Gondran, Michel (1986). *An Introduction to Expert Systems*. McGraw-Hill, Maidenhead.

Green, Shaw, Hurst, Leon, Nangle, Brenda, Cunningham, Pádraig, Somers, Fergal, and Evans, Evans (1997). Software agents: A review. Technical report, Trinity College, University of Dublin, Dublin.

Grosso, W., Eriksson, H., Fergerson, R., Gennari, Gennari, Tu, S., and Musen, M. (1999). Knowledge modeling at the millennium (the design and evolution of Protégé-2000). In *Proceedings of the Twelfth Workshop on Knowledge Acquisition, Modeling and Management.*

Gruber, T. R. (1992). Ontolingua: A mechanism to support portable ontologies.

Guessoum, Z. (2004). Adaptive agents and multiagent systems. *IEEE Distributed Systems Online*, 5.

Guha, Sudipto, Rastogi, Rajeev, and Shim, Kyuseok (1998). CURE: an efficient clustering algorithm for large databases. In *Proceedings of ACM SIGMOD International Conference on Management of Data*, pages 73–84.

Guha, Sudipto, Rastogi, Rajeev, and Shim, Kyuseok (2000). ROCK: A robust clustering algorithm for categorical attributes. *Information Systems*, 25(5):345–366.

Haeckel, S. H. and Nolan, R. (1994). *Managing by wire*, pages 122–132. Harvard Business Review.

Haefner, J. W. and Crist, T.O. (1994). Spatial model of movement and foraging in harvester ants (pogonomyrmex) (i): The roles of environment and seed dispersion. *Journal of Theoretical Biology*, 166:299–313.

Han, Jiawei and Fu, Yongjian (1994). Dynamic generation and refinement of concept hierarchies for knowledge discovery in databases. In *KDD Workshop*, pages 157–168.

Han, Jiawei and Kamber, Micheline (2001). *Data Mining: Concepts and Techniques.* Morgan Kaufmann Publishers.

Hayes-Roth, Frederick, Waterman, Donald A., and Lenat, Douglas B. (1983). Building expert systems. *Addison-Wesley Pub. Co., Reading, Mass, 444 p., 1983*, 1.

Herlocker, Jonathan L., Konstan, Joseph A., Borchers, Borchers, and Riedl, John (1999). An algorithmic framework for performing collaborative filtering. In *Proceedings of the 22nd Annual International ACM SIGIR Conference on Research and Development in Information Retrieval*, Theoretical Models, pages 230–237.

Hillbrand, E. and Stender, J. (1994). *Many-Agent simulation and Artificial Life.* IOS Press.

Hinneburg, A. and Keim, D. A. (1998). An efficient approach to clustering in large multimedia databases with noise. In *Proc. 4th Int. Conference on Knowledge Discovery in Databases (KDD'98), New York, NY*, pages 58–65.

Holland, J. (1995). *Hidden Order: How Adaptation Builds Complexity,*. Addison Wesley.

Holland, John H. (1975). *Adaptation in natural artificial systems.* University of Michigan Press, Ann Arbor.

Holland, John H. (1987). Genetic Algorithms and Classifier Systems: Foundations and Future Directions. In Grefenstette, John J., editor, *Proceedings of the 2nd International Conference on Genetic Algorithms (ICGA87)*, pages 82–89, Cambridge, MA. Lawrence Erlbaum Associates.

Holsapple, C. W. and Sena, M. P. (2004). Erp plans and decision-support benefits. *Decision Support Systems.*

Hong, T-P., Kuo, C-S., and Chi, S-C. (1999). Mining association rules from quantitative data. *Intelligent Data Analysis*, 3:363–376.

Hraber, Peter T., Forrest, Stephanie, and Jones, Jones (1997). The ecology of echo.

Huberman, Bernardo A. and Hogg, Tad (1988). The behavior of computational ecologies. In Huberman, B. A., editor, *The Ecology of Computation*, pages 77–176. North-Holland Publishing Company, Amsterdam.

Huhns, Michael N. and Singh, Munindar P., editors (1998). *Readings in Agents.* Morgan Kaufmann, San Francisco, CA, USA.

Huhns, Michael N. and Stephens, Larry M. (1999). Multiagent systems and societies of agents. In Weiss, Gerhard, editor, *Multiagent Systems: A Modern Approach to Distributed Artificial Intelligence*, chapter 2, pages 79–120. The MIT Press, Cambridge, MA, USA.

Hunt, E. B., Marin, J., and Stone, P. T. (1966). *Experiments in Induction.* Academic Press, New York.

Information Discovery Inc. (1999). *Datamines for Data Warehousing*, pages 1–15. Information Discovery Inc.

Jain, A. K., Murty, M. N., and Flynn, P. J. (1999). Data clustering: A survey. *ACM Computing Surveys*, 31:264–323.

Jennings, N. R. (1993). Commitements and conventions: The foundation of coordination in multi-agent systems. *The Knowledge Engineering Review*, 2:223–250.

Jennings, N. R., Corera, J., Laresgoiti, I., Mamdani, E. H., Perriolat, F., Sharek, P., and Varga, L. Z. (1996). Using archon to develop real-world dai applications for electricity transportation management and particle accelarator control. *IEEE Expert.*

Jennings, Nicholas R. and Wooldridge, Michael J., editors (1998). *Agent Technology: Foundations, Applications, and Markets.* Springer-Verlag: Heidelberg, Germany.

Kaelbling, L. P. and Rosenschein, S. J. (1990). *Action and planning in embedded agents.* The MIT Press.

Kargupta, H., Hamzaoglou, I., and Stafford, B. (1996). Padma: Parallel data mining agents for scalable text classification. In *Proceedings of the High Performance Computing.*

Kaufman, Leonard and Rousseeuw, Peter J. (1990). *Finding Groups in Data: An Introduction to Cluster Analysis.* John Wiley & Sons.

Kehagias, D., Symeonidis, A.L., and Mitkas, P.A. (2005). Designing pricing mechanisms for autonomous agents based on bid-forecasting. *Journal of Electronic Markets*, 15.

Kennedy, R. L., Lee, Y., Roy, B. Van, Reed, C. D., and Lippman, R. P. (1998). *Solving Data Mining Problems Using Pattern Recognition.* Prentice Hall PTR.

Kerber, R. (1992). ChiMerge: Discretization of numeric attributes. In Swartout, W., editor, *Proceedings of the 10th National Conference on Artificial Intelligence*, pages 123–128, San Jose, California. MIT Press.

Kero, B., Russell, L., Tsur, S., and Shen, W. M. (1995). An overview of data mining technologies. In *The KDD Workshop in the 4th International Conference on Deductive and Object-Oriented Databases.*

Kifer, Michael, Lausen, Georg, and Wu, James (1991). Logical foundations of object-oriented and frame-based languages. Technical Report 90/14, Department of Computer Science, State University of New York at Stony Brook, Stony Brook, NY 11794.

Knapik, Michael and Johnson, Jay (1998). *Developing intelligent agents for distributed systems: exploring architecture, technologies, & applications.* McGraw-Hill, Inc.

Kodratoff, Y. (1988). *Introduction to Machine Learning.* Pitman Publishing.

Konstan, Joseph A., Miller, Bradley N., Maltz, David, Herlocker, Jonathan L., Gordon, Lee R., and Riedl, John (1997). GroupLens: Applying collaborative filtering to Usenet news. *Communications of the ACM*, 40(3):77–87.

Koonce, David A., Fang, Cheng-Hung, and Tsai, Shi-Chi (1997). A data mining tool for learning from manufacturing systems. *Computer Industrial Engineering*, 33(3-4):27–30.

Krebs, F. and Bossel, H. (1997). Emergent value orientation in self-organization of an animat. *Ecological Modelling*, 96:143–164.

Kwon, O. B. and Lee, J. J. (2001). A multi agent intelligent system for efficient erp maintenance. *Expert Systems with Applications*, 21:191–202.

Labrou, Y., Finin, T., and Peng, Y. (1999). The current landscape of agent communication languages. *IEEE Intelligent Systems*, 14:45–52.

Labrou, Yannis and Finin, Tim (1998). Semantics for an agent communication language. In Singh, M. P., Rao, A., and Wooldridge, M. J., editors, *Intelligent Agents IV: Agent Theories, Architectures, and Languages*, volume 1365 of *Lecture Notes in Computer Science*, pages 209–214. Springer-Verlag, Heidelberg, Germany.

Langton, C. G. (1994). Personal communication.

Lawler, E. L., Lenstra, J. K., Kan, Kan, and Shmoys, D. B. (1985/1992). *The Traveling Salesman Problem.* Wiley.

Lee, C. Y. (1961). An algorithm for path connection and its applications. *IRE Transactions on Electronic Computers*, EC-10(3):346–365.

Lenat, Douglas B. (1995). A large-scale investment in knowledge infrastructure. *Communications of the ACM*, 38(11):33–38.

Levi, S. D., Kaminsky, P., and Levi, S. E. (2000). *Designing and managing the supply chain.* McGraw-Hill.

Liu, Bing, Chen, Shu, Hsu, Wynne, and Ma, Yiming (2001). Analyzing the subjective interestingness of association rules.

Looney, C. G. (1997). *Pattern Recognition Using Neural Networks: Theory and Algorithms for Engineers and Scientists.* Oxford University Press.

Luck, M., McBurney, P., and Preist, C. (2003). *Agent Technology: Enabling Next Generation Computing (A Roadmap for Agent Based Computing).* AgentLink.

Malone, T. W. (1998). Inventing the organizations of the twentieth first century: control, empowerment and information technology. In *Harvard Business Review Sept/Oct 1998*, pages 263–284. Harvard Business School Press.

Mangina, Eleni (2002). Review of software products for multi-agent systems. Technical report, AgentLink.org.

May, R. M. (1973). *Stability and Complexity in model ecosystems.* Princeton University Press.

McQueen, J. B. (1967). Some methods of classification and analysis of multivariate observations. In Cam, L. M. Le and Neyman, J., editors, *Proceedings of Fifth Berkeley Symposium on Mathematical Statistics and Probability*, pages 281–297.

Mehta, Manish, Agrawal, Rakesh, and Rissanen, Jorma (1996). SLIQ: A fast scalable classifier for data mining. In *Extending Database Technology*, pages 18–32.

Merriam-Webster (2000). [webster's dictionary].

Mitchell, Melanie (1996). *An Introduction to Genetic Algorithms.* Complex Adaptive Systems. MIT-Press, Cambridge.

Mitkas, P. A., Kehagias, D., Symeonidis, A. L., and Athanasiadis, I. (2003). A framework for constructing multi-agent applications and training intelligent agents. In *Proceedings of the 4th International Workshop on Agent-Oriented Software Engineering*, pages 1–16. Springer-Verlag.

Mitkas, P. A., Symeonidis, A. L., Kehagias, D., and Athanasiadis, I. (2002). An agent framework for dynamic agent retraining: Agent academy. In *Proceedings of the eBusiness and eWork 2002 (e2002) 12th annual conference and exhibition*, pages 757–764.

Mobasher, B. (1999). A web personalization engine based on user transaction clustering. In *Proceedings of the 9th Workshop on Information Technologies and Systems.*

Mobasher, B., Dai, H., Luo, T., Nakagawa, M., and Witshire, J. (2000a). Discovery of aggregate usage profiles for web personalization. In *Proceedings of the WebKDD Workshop.*

Mobasher, Bamshad, Cooley, Robert, and Srivastava, Jaideep (2000b). Automatic personalization based on Web usage mining. *Communications of the ACM*, 43(8):142-151.

Mobasher, Bamshad, Srivastava, Jaideep, and Cooley, Cooley (1999). Creating adaptive web sites through usage-based clustering of URLs.

Mohammadian, M. (2004). *Intelligent Agents for Data Mining and Information Retrieval.* Idea Group Inc.

Murrel, D. J., Travis, J. M. J., and Dytham, C. (2002). The evolution of dispersal distance in spatially-structured populations. *Oikos*, 97:229–236.

Nasraoui, O., Frigui, H., Joshi, A., and Krishnapuram, R. (1999). Mining web access logs using relational competitive fuzzy clustering. In *Proceedings of the Eight International Fuzzy Systems Association World Congress.*

Ng, R. T. and Han, J. (1994). Efficient and effective clustering methods for spatial data mining. In Bocca, Jorgeesh, Jarke, Matthias, and Zaniolo, Carlo, editors, *20th International Conference on Very Large Data Bases, September 12–15, 1994, Santiago, Chile proceedings*, pages 144–155, Los Altos, CA 94022, USA. Morgan Kaufmann Publishers.

Nwana, H. S. (1995). Software agents: An overview. *Knowledge Engineering Review*, 11(2):205–244.

Papoulis, A. (1984). *Probability, Random Variables, and Stochastic Processes.* EDITION:2nd; McGraw-Hill Book Company; New York, NY.

Parunak, Dyke, Van, H., Sauter, John, and Brueckner, Brueckner (2001). ERIM's approach to fine-grained agents.

Patel-Schneider, Peter F. (1998). Description-logic knowledge representation system specification from the KRSS group of the ARPA knowledge sharing effort.

Pecala, S. W. (1986). Neighborhood models of plant population dynamics. 2. multi-species models of annuals. *Theoretical Population Biology*, 29:262–292.

Peng, Y., Finin, T., Labrou, Y., Chu, B., Tolone, W., and Boughannam, A. (1999). A multi agent system for enterprise integration. *Applied Artificial Intelligence*, 13:39–63.

Perkowitz, Mike and Etzioni, Oren (1998). Adaptive web sites: Automatically synthesizing web pages. In *AAAI/IAAI*, pages 727–732.

Piatetsky-Shapiro, Gregory (1991). Discovery, analysis, and presentation of strong rules. In Piatetsky-Shapiro, Gregory and Frawley, William J., editors, *Knowledge Discovery in Databases*, pages 229–248. American Association for Artificial Intelligence, Menlo Park, California, U.S.A.

Pilot Software Inc. (1999). *White Paper: An introduction to Data Mining*, pages 1–12. Pilot Software Co.

Pyle, Dorian (1999). *Data Preparation for Data Mining.* Morgan Kaufmann, San Francisco.

Quinlan, J. R. (1986). Induction of decision trees. In Shavlik, Jude W. and Dietterich, Thomas G., editors, *Readings in Machine Learning.* Morgan Kaufmann.

Quinlan, J. R., editor (1987). *Applications of Expert Systems.* Addison Wesley, London.

Quinlan, J. R. (1992). *C4.5: Programs for Machine Learning.* Morgan Kaufmann.

Ray, Thomas S. (1992). An approach to the synthesis of life. In Langton, C. G., Tayler, C., Farmer, J. D., and Rasmussen, S., editors, *Artificial Life II*, pages 371–408. Addison-Wesley, Reading, MA.

Rosenschein, J. S. and Zlotkin, G. (1994). Designing conventions for automated negotiation. *AI Magazine*, pages 29–46.

Roure, D. De, Baker, M.A., Jennings, N.R., and Shadbolt, N.R. (2003). The evolution of the grid. Grid Computing: Making the Global Infrastructure a Reality.

Rousset, F. and Gandon, S. (2002). Evolution of the distribution of dispersal distance under distance-dependent cost of dispersal. *Journal of Evolutionary Biology*, 15:515–523.

Rust, R. T., Zeithaml, V. A., and Lemon, K. (2000). *Driving customer Equity: How customer lifetime value is reshaping corporate strategy.* The Free Press.

Rygielsky, C., Wang, J. C., and Yen, D. C. (2002). Data mining techniques for customer relationship management. *Technology in Society*, 24:483–502.

Schechter, Stuart, Krishnan, Murali, and Smith, Smith (1998). Using path profiles to predict HTTP requests. *Computer Networks and ISDN Systems*, 30(1–7):457–467.

Serugendo, G. Di Marzo and Romanovsky, A. (2002). Designing fault-tolerant mobile systems. In *Proceedings of the International Workshop on scientific engineering of Distributed Java applications*, volume 2604, pages 185–201. Springer-Verlag.

Shafer, John C., Agrawal, Rakesh, and Mehta, Manish (1996). SPRINT: A scalable parallel classifier for data mining. In Vijayaraman, T. M., Buchmann, Alejandro P., Mohan, C., and Sarda, Nandlal L., editors, *Proc. 22nd Int. Conf. Very Large Databases, VLDB*, pages 544–555. Morgan Kaufmann.

Shahabi, Cyrus, Zarkesh, Amir M., Adibi, Jafar, and Shah, Vishal (1997). Knowledge discovery from users web-page navigation. In *RIDE.*

Shapiro, J. (1999). *Bottom-up vs. top-down approaches to supply chain modeling*, pages 737–759. Kluwer.

Shen, Weiming, Norrie, Douglas H., and Barthes, Jean-Paul A. (2001). *Multi-agent Systems for Concurrent Intelligent Design and Manufacturing.* Taylor and Francis, London, UK.

Shortliffe, Edward (1976). *MYCIN: Computer-Based Medical Consultations.* American Elsevier.

Simon, H. (1996). *The Sciences of the Artificial.* MIT Press, MA.

Singh, M. P. (1997). Considerations on agent communication. In *FIPA Workshop.*

Spiliopoulou, Myra, Pohle, Carsten, and Faulstich, Lukas (1999). Improving the effectiveness of a web site with web usage mining. In *WEBKDD*, pages 142–162.

Stolfo, Salvatore, Prodromidis, Andreas L., Tselepis, Shelley, Lee, Wenke, Fan, Dave W., and Chan, Philip K. (1997). JAM: Java agents for meta-learning over distributed databases. In Heckerman, David, Mannila, Heikki, Pregibon, Daryl, and Uthurusamy, Ramasamy, editors, *Proceedings of the Third International Conference on Knowledge Discovery and Data Mining (KDD-97)*, page 74. AAAI Press.

Sycara, K., Paolucci, M., Ankolekar, A., and Srinivasan, N. (2003). Automated discovery, interaction and composition of semantic web services. *Journal of Web Semantics*, 1:27–46.

Sycara, K., Wooldridge, M. J., and Jennings, N.R. (1998). A roadmap of agent research and development. *International Journal of Autonomous Agents and Multi-Agent Systems*, 1:7–38.

Symeonidis, A. L., Kehagias, D., and Mitkas, P. A. (2003). Intelligent policy recommendations on enterprise resource planning by the use of agent technology and data mining techniques. *Expert Systems with Applications*, 25:589–602.

Talavera, L. and Cortes, U. (1997). Inductive hypothesis validation and bias selection in unsupervised learning. In *Proceedings of the 4th European Symposium on the Validation and Verification of Knowledge Based Systems*, pages 169–179.

The FIPA Foundations (2000). Fipa-sl specifications, 2000, fipa sl content language specification. Technical report, The FIPA Consortium.

The FIPA Foundations (2003). Foundation for intelligent physical agents specifications. Technical report, The FIPA Consortium.

Traum, David R. (1999). Speech acts for dialogue agents.

Tschudin, Christian (1995). OO-Agents and Messengers.

Turney, Peter D. (2002). Robust classification with context-sensitive features. In *Proceedings of the Sixth International Conference on Industrial and Engineering Applications of Artificial Intelligence and Expert Systems.*

Wang, H., Huang, J.Z., Qu, Y., and Xie, J. (2004). Web services: problems and future directions. *Journal of Web Semantics*, 1:309–320.

Weiss, G. (1999). *Multiagent Systems: A Modern Approach to Artificial Intelligence.* The MIT Press.

Weiss, Sholom M. and Indurkhya, Nitin (1998). *Predictive data mining: a practical guide.* Morgan Kaufmann Publishers Inc.

Weiss, Sholom M. and Kulikowski, Casimir A. (1991). *Computer Systems That Learn: Classification and Prediction Methods from Statistics, Neural Nets, Machine Learning, and Expert Systems.* Morgan Kaufmann, San Mateo, California.

Werner, Gregory M. and Dyer, Michael G. (1994). Bioland: A massively parallel simulation environment for evolving distributed forms for intelligent behavior. In Kitano, Hitoaki and Hendler, James A., editors, *Massively Parallel Artificial Intelligence*, pages 316–349. AAAI Press/MIT Press.

Westerberg, L. and Wennergren, U. (2003). Predicting the spatial distribution of a population in a heterogeneous landscape. *Ecological Modelling*, 166:53–65.

White, J. E. (1994). Telescript technology: The foundation for the electronic marketplace. White paper, General Magic, Inc.

Wilson, S. W. (1987). Classifier systems and the animat problem. *Machine Learning*, 2:199–228.

Wilson, S. W. and Goldberg, D. E. (1989). A critical review of classifier systems. In *Proceedings of the third international conference on Genetic algorithms*, pages 244–255. Morgan Kaufmann.

Wilson, Stewart W. (1990). The Animat Path to AI. In Meyer, J. A. and Wilson, S. W., editors, *From Animals to Animats 1. Proceedings of the First International Conference on Simulation of Adaptive Behavior (SAB90)*, pages 15–21. A Bradford Book. MIT Press.

Witten, Ian H. and Frank, Eibe (2000). *Data Mining: Practical Machine Learning Tools and Techniques with Java Implementations.* Morgan Kaufman.

Wooldridge, M. (1997). Agent-based software engineering. *IEE Proceedings Software Engineering*, 144(1):26–37.

Wooldridge, Michael (1999). Intelligent agents. In Weiss, Gerhard, editor, *Multiagent Systems: A Modern Approach to Distributed Artificial Intelligence*, chapter 1, pages 27–78. The MIT Press, Cambridge, MA, USA.

Wooldridge, Michael and Jennings, Nicholas R. (1995). Intelligent agents: Theory and practice. *Knowledge Engineering Review*, 10(2):115–152.

Worley, J. H., Castillo, G. R., Geneste, L., and Grabot, B. (2002). Adding decision support to workflow systems by reusable standard software components. *Computers in Industry*, 49:123–140.

Xingdong, Wu (1995). *Knowledge Acquisition from Databases.* Ablex Publishing Corp., Greenwich, USA.

Yaeger, Larry (1994). Computational genetics, physiology, metabolism, neural systems, learning, vision, and behavior or PolyWorld: Life in a new context. In Langton, Christopher G., editor, *Proceedings of the Workshop on Artificial Life*, volume 17 of *Sante Fe Institute Studies in the Sciences of Complexity*, pages 263–298, Reading, MA, USA. Addison-Wesley.

Zaïane, Osmar R., Xin, Man, and Han, Jiawei (1998). Discovering web access patterns and trends by applying OLAP and data mining technology on web logs. In *Advances in Digital Libraries*, pages 19–29.

Zhang, Chengqi and Zhang, Shichao (2002). *Association rule mining: models and algorithms.* Springer-Verlag New York, Inc.

Zhang, Tian, Ramakrishnan, Raghu, and Livny, Miron (1996). BIRCH: an efficient data clustering method for very large databases. In *Proceedings of the 1996 ACM SIGMOD International Conference on Management of Data*, pages 103–114.

Zhang, Z., Zhang, C., and Zhang, S. (2003). An agent-based hybrid framework for database mining. *Applied Artificial Intelligence*, 17:383–398.

Index

About the Authors

Andreas L. Symeonidis received his Diploma and PhD from the Department of Electrical and Computer Engineering at the Aristotle University of Thessaloniki in 1999 and 2004, respectively. Currently, he is a Postdoctoral Research Associate with the university. His research interests include software agents, data mining and knowledge extraction, intelligent systems, and evolutionary computing (e-mail: asymeon@iti.gr).

Pericles A. Mitkas received his Diploma of Electrical Engineering from Aristotle University of Thessaloniki in 1985 and an MSc and PhD in Computer Engineering from Syracuse University, USA, in 1987 and 1990, respectively. He is currently an Associate Professor with the Department of Electrical and Computer Engineering at the Aristotle University of Thessaloniki, Greece. He is also a faculty affiliate of the Informatics and Telematics Institute of the Center for Research and Technology - Hellas (CERTH). His research interests include databases and knowledge bases, data mining, software agents, enviromatics and bioinformatics. Dr Mitkas is a senior member of the IEEE Computer Society. His work has been published in over 120 papers, book chapters, and conference publications (e-mail: mitkas@eng.auth.gr).

Printed by Publishers' Graphics LLC